GEOGRAPHERS

Biobibliographical Studies

VOLUME 3

Edited by

T. W. Freeman and Philippe Pinchemel

on behalf of the

Commission on the History of Geographical Thought

of the International Geographical Union

and the International Union of the History

and Philosophy of Science

MANSELL 1979

© 1979 International Geographical Union

Mansell Publishing,
3 Bloomsbury Place, London WC1A 2QA
First published 1979

This volume forms part of the series *Studies in the
History of Geography* planned by the International
Geographical Union, Commission on the History of
Geographical Thought. *Chairman,* Professor Philippe
Pinchemel, U.E.R. de Géographie, Université de Paris,
191 rue Saint Jacques, 75005, Paris. *Secretary,*
Professor T.W. Freeman, 1 Thurston Close, Abingdon
OX14 5RD. *Ordinary Members:* Professor Vladimir
Annenkov, Geographical Institute, Academy of Sciences,
Staromonetny per 29, Moscow V-17; Professor Jozef
Babicz, Institut d'Histoire des Sciences et des Techniques,
Polska Akademia Nauk, Nowy Swiat 72, Warsaw; Professor
George Kish, Department of Geography, University of
Michigan, Ann Arbor, Michigan 48104; Professor Jozef
Schmithüsen, Universität des Saarlandes, Geographisches
Institut, Universität Bau 11 IV, D-66, Saarbrücken
15, West Germany; Professor Ichiro Suitsu, Institute
of Geography, Kyoto University, Yoshida, Sakyo-Ku-
Kyoto, Japan; *Honorary members:* Mr Gerald
Crone, 34 Cleveland Road, Ealing, London W13; Professor
Robert E. Dickinson, 636 West Roller Coaster Road,
Tucson, Arizona 85704; Professor R. Hartshorne,
Department of Geography, University of Wisconsin,
Madison, Wisconsin 53706.

International Standard Book Number 0 7201 0927 2

International Standard Serial Number 0308-6992

Enquiries concerning publications listed in the 'Bibliography
and Sources' sections of the biobibliographies should
be sent to the relevant publisher or journal, and *not*
to Mansell.

British Library Cataloguing in Publication Data
Geographers: biobibliographical studies. -
(Studies in the history of geography).
  Vol. 3
  1. Geographers - Biography
  I. Freeman, Thomas Walter II. Pinchemel, Philippe
  III. Commission on the History of Geographical Thought
  IV. International Union of the History and Philosophy
  of Science
  V. Series
  910'.92'2   G67

  ISBN 0-7201-0927-2
  ISSN 0308-6992
Printed in Great Britain by photolithography and bound
at The Scolar Press, Ilkley, West Yorkshire.

Contents

Introduction

Before the first volume of *Geographers: Biobibliographical Studies* was published, there was a discussion of the order in which the papers should appear. The biobibliographies could have been arranged by countries and by dates of birth or death, or alphabetically by surnames. The latter alternative was considered to be the safest, if the least inspired, given the Editors' reliance on individual contributors of many nationalities. But if it had been feasible to choose a birth and death arrangement with the geographers geographically grouped together some interesting comparisons and contrasts would have emerged. In this volume Finland, France and Germany, Great Britain and the U.S.A. are represented by two or more geographers; whereas the remaining three countries, Australia, Romania and the U.S.S.R. have only one geographer each — although Australia's Griffith Taylor was such a dramatic figure that he could almost be said to be equivalent to two geographers.

Finland, most individual of countries, had a fine advocate for its identity in Topelius during the nineteenth century. Granö as a geographer was at once practical and inspired. His ideas, notably on perception, were novel and were to emerge as such later on. With other Finnish geographers of his time he was a keen student of the national identity of his country and of its detailed regionalization.

France is represented by five geographers. Of these Bernard belongs to North Africa and the colonial era. D'Abbadie, the earliest, was a traveller in Ethiopia and one of the many people obsessed with the source of the river Nile. Reclus, sometimes called an encyclopaedist but better regarded as the compiler of a *Universal Geography*, had periods of exile because of his anarchist views, which were by no means unique among the intelligentsia at that time. Ancel and Arbos, born in the same year, had contrasting lives. Ancel was drawn firmly into political geography by the First World War and showed his concern, shared by many of his contemporaries, for the future of Europe, and particularly of France. Arbos's life passed in a more peaceful way: he loved the Alps with their pastoral life, and the Auvergne, and spent his working years mainly at the University of Clermont-Ferrand.

Germany has the classic figures of Melanchthon and Münster, looking at geography in terms of a world explicable only by philosophy and theology. They asked the fundamental questions only sincere and courageous people dare to ask at a time when it was not always safe to do so.

Of the four British geographers included, three were born in Scotland. The two Geikies may be regarded as the greater Archibald and the lesser James, the former of whom settled in England, but the latter did not and apparently never wished to do so. Both were known mainly as geologists and physiographers but were anxious that geography should be well taught in schools — though they were less enthusiastic about its growth in the universities. However James was a devoted worker for the Royal Scottish Geographical Society and as honorary editor worked with Marion Newbigin to make its journal, the *Scottish Geographical Magazine*, internationally known and respected. Herbertson came to geography through the famous Bartholomew map firm in Edinburgh, through writing and editing geography texts for schools, and

through lecturing to adult audiences in universities
and outside them.   In time he was appointed lecturer
at Oxford University where his teaching, editing,
writing and other activities continued to his death.
Through Herbertson, and the more charismatic figure of
H. J. Mackinder, the regional tradition was firmly
established in British geography and Gilbert, trained
as an historian, made it the foundation of later
developments at Oxford, in which a concern for the
preservation of the landscape was linked with an
understanding of current planning problems.

All four Americans who appear had the energy and
strength to carry out fieldwork, in Platt's case
abroad as well as in his homeland.   Powell stands out
as a fine geomorphologist led by experience to realize
that the settlement of pioneering areas was a matter
demanding careful scientific study.   That he was
right has since been abundantly demonstrated but
politicians took a different view and he ran into
difficulties.   Atwood and Shaler also knew American
environments well from fieldwork, though their contribution to geography was made primarily as edu-
cationalists.   All four did much to impart a stronger
emphasis to physiography in American geography, and
the idea of environmental conservation appears in
their varied work long before it became a fashionable
cause.

Fieldwork was also the basis of the contribution
to geography made by the Russian Tillo, with a military background, who gave fine service to cartography.

Australia had the vivid figure of Griffith Taylor,
maligned by politicians for what they eventually had
to accept as a reasonable appraisal of the possibilities of agricultural settlement.   Taylor belonged
to Canada and the U.S.A. for part of his life, and his
fieldwork, some of it necessarily rapid, covered a
multitude of countries.   He undoubtedly enjoyed the
cut and thrust of controversy.   A far quieter figure
who had only a short life, Dimitrescu-Aldem of
Romania, shared the wish of numerous colleagues to be
a complete geographer, though his main interest was in
geomorphology, and like his contemporaries in other
countries he did much to advance geography teaching at
all levels.

So diverse a score of geographers, in such different times, places and circumstances, might at first
seem to have no relationship but in reality all made
some offering to the development of geographical
thought, the strength and the weakness of which is
only revealed in the fullness of time.   Not always is
the successful figure able to assume that the judgement of posterity will be what he might wish.

T.W. Freeman

Note:

Intending authors are asked to write to Professor
T.W. Freeman, 1 Thurston Close, Abingdon, OX14
5RD, England, who will send a note of information
for authors of biobibliographical studies.

List of Abbreviations

Abbreviations have been adopted from *British Standard 4148: Part 2, 1975, Word-abbreviation list,* and refer to abbreviations in both the bibliographical references and chronological tables.

Abh. Gesell. Wiss. Göttingen, Philol. Hist. Kl.
Abhandlung der Gesellschaft der Wissenschaften zu Göttingen, Philologisch Historische Klasse

Abh. K. Sächsischen Akad. Wiss., Philol. Hist. Kl.
Abhandlung der Königlich Sächsischen Akademie der Wissenschaften, Philologisch Historische Klasse

Acta Geogr. Acta Geographica

Allg. Dtsch. Biogr. Allgemeine Deutsche Biographie

Am. Anthropol. American Anthropology

Am. Bar. Assoc. Rep. ... Ann. Meet. American Bar Association Report ... Annual Meeting

Am. Geogr. Soc. Spec. Publ. American Geographical Society Special Publication

Am. Geol. American Geologist

Am. J. Sci. Arts. American Journal of Science and Arts

Am. J. Sociol. American Journal of Sociology

Ann. Assoc. Am. Geogr. Annals of the Association of American Geographers

Ann. de la Soc. Sci. de Bruxelles Annales de la Société Scientifiques de Bruxelles

Ann. Géogr. Annales de Géographie

Ann. Hist. Soc. Annales d'Histoire Sociale

Ann. Univ. Grenoble Annales de l'Université de Grenoble

Arch. Kult.-gesch. Archiv für Kulturgeschichte

Assoc. Am. Geogr. Association of American Geographers

Atl. Mon. Atlantic Monthly

Atti del Terzo Congr. Geogr. Int. Atti del Terzo Congresso Geografico Internazionale

Austr. Geogr. Australian Geographer

Austr. Geogr. Stud. Australian Geographical Studies

Australasian Assoc. Adv. Sci. Rep. Australasian Association for the Advancement of Science, Reports

Basler Beit. Geschi.-Wiss. Basler Beiträge Geschtswissenchaft

Beitr. Ingelheimer Gesch. Beiträge zur Ingelheimer Geschichte

Bibl. Sociol. Bibliothèque Sociologique

Boll. Soc. Geogr. Ital. Bollettino Societa Geografica Italiana

Brit. Assoc. Adv. Sci., Rep. Ann. Meet. British Association for the Advancement of Science, Report of Annual Meeting

Bul. Soc. (regale) Rom. De Geogr. Buletinul Societatii (regale) Române de Geografie

Bull. Assoc. Geogr. Fr. Bulletin de l'Association des Géographes Françaises

Bull Bur. Meteorol. Bulletin of the Bureau of Meteorology

Bull. de la Soc. Ramond Bulletin de la Société Ramond

Bull de l'Ouest Bulletin de l'Ouest

Bull. Geogr. Soc. Chicago Bulletin of the Geographical Society of Chicago

Bull. Geol. Soc. Am.   Bulletin of the Geological Society of America

Bull. Philos. Soc. Washington   Bulletin of the Philosophical Society of Washington

Bull. Sect. Géogr. Com. Trav. Hist. Sci.   Bulletin de la Section de Géographie du Comité des Travaux Historiques et Scientifiques

Bull. Soc. Belge d'Étud. Géogr.   Bulletin de la Société Belges d'Études Géographiques

Bull. Soc. d'Émulation du Bourbonnais   Bulletin de la Société d'Émulation du Bourbonnais

Bull. Soc. Geogr.   Bulletin de la Société de Géographie

Bull. Soc. Géogr. et Archéol. d'Oran   Bulletin de la Société Géographique et Archéologique d'Oran

C.R. Acad. des Sci.   Comptes-Rendus, Académie des Sciences

C.R. Congr. Int. Géogr.   Comptes-Rendus, Congrès Internationale de Géographie

Can. Banker   Canadian Banker

Can. Geogr.   Canadian Geographer

Can Inst. Int. Aff.   Canadian Institute of International Affairs

Can J. Econ, Polit. Sci.   Canadian Journal of Economic and Political Science

Commonw. Bur. Meteorol.   Commonwealth Bureau of Meteorology

Commonw. Yearb.   Commonwealth Yearbook

Congr. de l'Afr. du Nord, C.R. des Trav.   Congrès de L'Afrique du Nord, Comptes-Rendus des Travaux

Congr. Sci. de France   Congrès Scientifique de France

Corp. Reform.   Corpus Reformatorum

Dict. Am. Biog.   Dictionary of American Biography

Dict. Biog. Fr.   Dictionnaire de Biographie Française

Dict. Sci. Biog.   Dictionary of Scientific Biography

Die Erde (Z. Gesell. Erdk.)   Die Erde (Zeitschrift der Gesellschaft für Erdkunde)

Econ. Geogr.   Economic Geography

Edinburgh Rev.   Edinburgh Review

Essex Inst. Hist. Collections   Essex Institute Historical Collections

Fortn. Rev.   Fortnightly Review

Fortschr. Geol. u. Palaeont.   Fortschrift für Geologie und Palaeontologie

Geogr.   Geography

Geogr. Anzeiger   Geographisches Anzeiger

Geogr. Bull.   Geographical Bulletin

Geogr. J.   Geographical Journal

Geogr. Mag.   Geographical Magazine

Geogr. Rev.   Geographical Review

Geogr. Rundschau   Geographische Rundschau

Geogr. Soc.   Geographical Society

Geogr. Taschenb.   Geographisches Taschenbuch

Geogr. Teach.   Geographical Teacher

Geogr. Z.   Geographische Zeitung

Geol. Mag.   Geological Magazine

Geol. Soc. Am. Bull.   Geological Society of America Bulletin

Gutenberg-Jahrb.   Gutenberg-Jahrbuch

Haslemere Nat. Hist. Soc. Sci. Pap.   Haslemere Natural History Society Scientific Papers

Hum. Biol.   Human Biology

I.G.U.   International Geographical Union

Illinois State Board of Educ. Proc.   Illinois State Board of Education Proceedings

Inst. de France Acad. Sci.   Institut de France, Académie des Sciences

Int. Geogr. Congr.   International Geographical Congress

Int. Geol. Congr.   International Geological Congress

Int. J.   International Journal

Int. Mon.   International Monthly

Izv. Akad. Nauk SSSR Ser. Geol.   Izvestiya Nauk SSSR seriya Geolica

Izv. Zapad. Sibirsk. Russ. Geogr. Obshch.   Izvestiya Zapadnova Sibirskovo Russkogo Geograficheskogo Obshchestva

J. Am. Geogr. Soc.   Journal of the American Geographical Society

J. Asiat.   Journal Asiatique

J. Geogr.   Journal of Geography

J. Geol.   Journal of Geology

J. R. Geogr. Soc.   Journal of the Royal Geographical Society

J. School Geogr.   Journal of School Geography

J. Scott. Meteor. Soc.   Journal of the Scottish Meteorological Society

J. Soc. Bibliogr. Nat. Hist.   Journal of the Society for the Bibliography of Natural History

J. Soc. Finno-Ougrienne   Journal de la Société Finno-Ougrienne

Jahresber. Ger. Natl.-Mus.   Jahresbericht des Germanischen National-Museum

La Géogr.   *La Géographie*

La Soc. Nouv.   La Société Nouvelle

Meteorol. Collected Works Acad. Sci.   Meteorological Collected Works of the Academy of Sciences

N. Am. Rev.   North American Review

Natl. Geogr. Mag.   National Geographic Magazine

Natl. Geogr. Monogr.   National Geographic Monographs

Natl. Geogr. Soc. Monogr.   National Geographic Society Monograph

Neue Z. für Syst. Theologie und Religionsphilos.   Neue Zeitschrift für Systematische Theologie und Religionsphilosophe

Nouv. Ann. des Voyages   Nouvelles Annales des Voyages

Pan-Am. Geol.   Pan-American Geologist

Pan. Am. Inst.   Pan-American Institute

Petermanns Geogr. Mitt.   Petermanns Geographische Mitteilungen

Philos. Nat.   Philosophia Naturalis

Pop. Sci. Mon.   Popular Science Monthly

Proc. All-Union Geogr. Soc.   Proceedings of the All-Union Geographical Society

Proc. Am. Assoc. Advance. Sci.   Proceedings of the American Association for the Advancement of Science

Proc. Biol. Soc. Washington   Proceedings of the Biological Society of Washington

Proc. Geol. Assoc.   Proceedings of the Geological Association

Proc. Geol. Soc. Am.   Proceedings of the Geological Society of America

Proc. ... Int. Geogr. Congr.   Proceedings of the... International Geographical Congress

Proc. Pan-Pac. Sci. Congr.   Proceedings of the Pan-Pacific Science Congress

Proc. R. Geogr. Soc.   Proceedings of the Royal Geographical Society

Proc. R. Soc.   Proceedings of the Royal Society

Proc. R. Soc. Edinburgh   Proceedings of the Royal

Society of Edinburgh
Proc. Russ. Geogr. Soc. Proceedings of the Russian Geographical Society
Proc. ... Sess, Inst. Int. Relations Proceedings of the ... Session of the Institute of International Relations
Publ. Clark Univ. Lib. Publications of the Clark University Library
Publ. Inst. Univ. Tartuensis Geogr. Publicationes Instituti Universitatis Tartuensis Geographici
Publ. Sem. Univ. Tartuensis Oecon.-Geogr. Publicationes Seminarii Universitatis Tartuensis Oeconomico-Geographici
Q. J. Geol. Soc. Quarterly Journal of the Geological Society
Q. J. R. Meteorol. Soc. Quarterly Journal of the Royal Meteorological Society
Quell. Forsch. Reformationsgeschichte Quellen und Forschungen zur Reformationsgeschichte
Quest Dipl. et Colon. Questions Diplomatiques et Coloniales
Rensiegn. Colon. et Doc. Com. Afr. Fr. et Com. Maroc Reseignements Coloniaux et Documents du Comité de l'Afrique Française et Comité du Maroc
Rep. Russ. Geogr. Soc. Report of the Russian Geographical Society
Rev. Afr. Revue Africaine
Rev. d'Auvergne Revue d'Auvergne
Rev. des Questions Sci. Revue des Questions Scientifiques
Rev. Deux Mondes Revue des Deux Mondes
Rev. Fr. Revue Française
Rev. Géogr. Revue Géographique
Rev. Géogr. Alp. Revue de Géographie Alpine
Rev. Geogr. Inst. Pan-Am. Geogr. Hist. Revista Geographica del Instituto Pan-Americano de Geographica e Historia
Rev. Géogr. Lyon Revue de Géographie de Lyon
Rev. Germ. Revue Germanique
Rev. Hist. Revista de Historia
Rev. Hist. du Sud-Est Eur. Revue Historique du Sud-Est Européen
Rev. Numismatique, nouv. sér. Revue Numismatique, nouvelle série
Rev. Polit. et Parlementaire Revue Politique et Parlementaire
Rheinische Bl. Erzich. Unterricht. Rheinische Blatter für Erziehung und Unterrichtung
Riv. Geogr. Ital. Rivista Geografica Italiana
Sch. and Soc. School and Society
Schweiz. Z. Gesch. Schweizerische Zeitschrift für Geschichte
Sci. Mon. Scientific Monthly
Scott. Geogr. Mag. Scottish Geographical Magazine
Scribner's Mag. Scribner's Magazine
Sitzungber. Finn. Acad. Wissenschaften Sitzungsberichte der Finnischen Akademie der Wissenschaften
Smithson Inst. Bur. Ethnol. ... Ann. Rep. Smithsonian Institution Bureau of Ethnology Annual Report
Smithson. Rep. Smithsonian Institution Annual Report
Sudhoffs Arch. Sudhoffs Archiv
Sven. Geogr. Årsbok Svensk Geografiska Årsbok

Temps. Nouv. Temps Nouvelles
Trans. Am. Inst. Min. Eng. Transactions of the American Institute of Mining Engineers
Trans Anthropol. Soc. Washington Transactions of the Anthropological Society of Washington
Trans ... Congr. Russ. Nat. Physicians Transactions of the ... Congress of Russian Naturalists and Physicians
Trans. Edinburgh Geol. Soc. Transactions of the Edinburgh Geological Society
Trans. Geol. Soc. Glasgow Transactions of the Geological Society of Glasgow
Trans. Illinois State Acad. Sci. Transactions of the Illinois State Academy of Science
Trans. Inst. Br. Geogr. Transactions of the Institute of British Geographers
Trans. Military Topog. Dept. Transactions of the Military Topographical Department
Trans. Moscow Inst. Geod. Aerial Photogr. Cartogr. Transactions of the Moscow Institute for Geodesy, Aerial Photography and Cartography
Trans. Orenburg Dept. Russ. Geogr. Soc. Transactions of the Orenburg Department of the Russian Geographical Society
Trans R. Soc. Edinburgh Transactions of the Royal Society of Edinburgh
Trans. Southeast Union Sci. Soc. Transactions of the Southeast Union of Scientific Societies
Trav. Inst. Géogr. Alp. Travaux de l'Institut de Géographie Alpine
U. S. Geol Surv., ... Ann. Rep. United States Geolical Survey, ... Annual Report
U. S. Geol. Surv. Prof. Pap. United States Geological Survey Professional Papers
Univ. Chicago Dept. Geog. Res. Pap. University of Chicago Department of Geography Research Paper
Verh. Dtsch. Geogr. Verhandlungen des Deutschen Geographentage
Wisconsin Geol. and Nat. Hist. Surv. Wisconsin Geological and Natural History Survey
Yearb. of Austr. Acad. Sci. Yearbook of the Australian Academy of Science
Z. für Geopolitik Zeitschrift für Geopolitik
Z. Gesell. Erdk. Zeitschrift der Gesellschaft für Erdkunde
Z. Vulkanol. Zeitschrift für Vulkanologie

Jacques Ancel
1882–1943

ROBERT SPECKLIN

Towards the end of 1830 Jacob Bernard, a horse dealer from Foussemagne, settled in Besançon and worked for the postmaster, Simon Ancel. The two families intermarried and in 1880 Tristan Bernard went to Paris and entered the Lycée Fontanes. Two years later, Jacques Ancel was born near Paris, on 22 July 1882. Thirty years later Ancel reviewed his life as 'study in Paris, first at the Lycée Condorcet and then at the Sorbonne. Licence ès lettres, teacher from 1905 to 1907 at schools in Vannes and Péronne, *agrégé* in history and geography in 1908 and then teacher at the Chaptal school in Paris'. So far it seemed that he would have a useful career as a schoolmaster but the stimulus of war brought a dramatic change in his life.

1. EDUCATION, LIFE AND WORK

Called up on 3 August 1914 as a soldier in the 167th Infantry Regiment, he was wounded at Bois le Prêtre, Lorraine, on 2 November 1914 and remained in hospital until 1 February 1915. Promoted rapidly to corporal and then to sergeant he was wounded again on 16 March 1915, also at Bois le Prêtre. He became a second lieutenant attached to the 21st Infantry Regiment on 29 December 1915 and fought in the Battle of Verdun from 4 to 12 March 1916, after which he was decorated with the Croix de Chevalier de la Légion d'honneur and the Croix de Guerre and was mentioned in despatches on 2 April 1916. He was then posted to the eastern front on 31 August 1916 and attached to the General Staff of the Eastern Army, then to the 57th and later the 17th colonial divisions of the allied eastern armies. He became head of the political department of the General Staff at Salonica from 3 February to 16 November 1918 under the command of generals Guillaumat and Franchet d'Esperey. Promoted to lieutenant in 1917 and captain in 1926 he became an Officer de la Légion d'honneur (military division) on 13 March 1933. Through his military service Ancel discovered the Balkans, to which most of his published work was given from 1919. His writings were welcomed and in 1926 he left the Chaptal school, to which he had returned after the war, and spent the rest of his working life at the Institut des Hautes Études Internationales of the University of Paris. In the following year he became closely associated with the Carnegie Centre Européen de la Dotation, which was formed in the Institute and under an American principal, Earle B. Babcock (1881-1935), studied ways of finding international peace.

Ancel owed his success to the clarity, convincing content and precision of his published work. He was determined to be clear and accurate and in 1932 Jules Sion said of his thesis that it was 'precise, colourful and living'. Later, in 1945, Lucien Febvre spoke of the immense vitality of the writing of Ancel, and of its warm glow combined with scholarly precision. Having a late start (he was twenty-six at the time of his *agrégation*), he became expert in drafting a paper or a lecture on a logical structure of two or perhaps three sections. In his work on the frontiers of Romania, for example, he noted that there were three mountain borders, of Moldavia, Walachia and Transylvania. He subdivided the Moldavian frontier into the beech forests of Bukovina, the *bocage* of Besserabia and the steppes of Bugeac. This method made it possible for him to give a vivid impression of the land-scape, acquired (as the Romanian geographer Iorga (1887-1946) observed) by long and patient field study. He found that Macedonia was also divisible into three sections but added a special chapter on Salonica.   In writing his *Manuel historique de la question d'Orient* he had to include twelve chronological chapters. Logical arrangement was also notable in his *Confins occidentaux* of German Europe, treated in twelve chap-ters, and he also found it possible to divide Germany itself into a dozen parts.

Obviously a worker, Ancel was also a fighter. That his military career was marked by valour is clear but here the interest lies in his academic determin-ation.   This was apparent in his work on Macedonia, which in the preface to his thesis he described as 'perhaps the most difficult problem in Europe, to some apparently impossible of solution'.   He adds that in his work he had considered every aspect of Macedonia and had not allowed his curiosity to be daunted by any obstacle.   In the thesis there is a map showing his widespread travels in Macedonia.   He stated quite plainly that he could never be convinced by the argu-ments of the pseudo-science of geopolitics and his best book, *Frontières*, published in three parts in 1938, was a reply to the *Grenzen (Frontiers)* of Karl Haushofer.

Ancel's last years were clouded by war circum-stances.   He was arrested on 12 December 1941, interned at Compiègne and released three months later on 12 March 1942.   In the camp at Compiègne he met his relative, Jean Jacques, son of Tristan Bernard. Lucien Febvre drew attention to his courage in facing the German invaders of France.

2. *SCIENTIFIC IDEAS AND GEOGRAPHICAL THOUGHT*

The main scientific contribution of Ancel was his thesis of 1930 on Macedonia, which was developed from his earlier work (of considerable erudition) on his own observations as a soldier, published in 1921. Most of his other work was of a more popular charac-ter, in the best sense of the term.   He gave a mas-terly presentation of complex issues and his texts for students were greatly appreciated. These included his book on *La question d'Orient (The Eastern Question)*, a subject of such complexity that it was likely to strike terror into the heart of any student; a manual on the diplomatic history of Europe edited by Henri Hauser in which most of the chapters were by Jacques Ancel and Pierre Renouvin; and also his own *Manuel géographique de politique européenne*.   Unfor-tunately he was unable to complete this work, which began with Central Europe: he finished the work on the western boundaries of Germany but not the work on the eastern boundaries, with Poland.   In all these learned works Jacques Ancel included material from his lectures, travels and discussions with others.

In Ancel's view geography was closely related to history.   Writing in 1922 he said that the relevance of geography was increasingly appreciated and in his book of 1923 on *La question d'Orient* he included a preliminary chapter on 'the geographical basis' in which 'areas of circulation' followed 'domains of civilization'. His study of the political geography of the Baltic (1931) deals solely with geographi-cal 'bases' or 'conditions'.   Conviction on the significance of geographical factors appeared to grow for in 1936 he said that the need of the day was to think in geographical terms.   According to his rela-tive and fellow-prisoner at Compiègne, Jean Jacques Bernard, in discussion Ancel applied himself to dis-cerning all the 'moral and material, psychological, historical, economic and climatic' influences which led people living within the same frontiers to be woven into a nation.

Ancel made it abundantly clear that geography should never be used to make a hasty synthesis: in a review of the work by Jean Brunhes and Camille Vallaux on the *Géographie de l'histoire* in 1922 he said that the analysis they gave was incomplete and therefore a synthesis could only be provisional. At the end of his 1930 thesis he explains that his work was purely analytical and that he was completely opposed to 'the ambition of certain German professors, centred around the *Zeitschrift für Geopolitik* to sweep forth into vast syntheses when the analyses are not adequately discerned and explained'.   In 1936 he said that he could only write a synthesis on political geography on a basis of rigorous analysis and he described his own *Manuel* as analytical and strictly regional.   This same conception of geography, strengthened by his work on the Balkans, is seen in his posthumous work on Germany in twelve chapters.   Only 35 pages are given to a general review compared with 225 on the old provinces: Saxony, Bavaria, Silesia, Brandenburg, East Prussia and even, as a concession to the circumstances of the time, Danzig.   This book does not provide a concise synthetic treatment of Germany as an entity, though nobody was more aware than Ancel that Germany was 'one and indivisible'.

Of his debt to other French geographers Ancel had no doubts at all.   He spoke of his 1926 book on the Balkans as an application 'in a defined area of teach-ing ... [by] ... the undoubted master of geography in France, [Vidal de la Blache], who had written little but though much' and to whom 'none of those who had heard him speak, working with his inspiration, could ever cease to be indebted'.   He noted in his preface to the 1936 book *(Géopolitique)* that the same spirit was apparent in the work of Albert Demangeon and on Central Europe he spoke of the work of Emmanuel de Martonne as a kind of 'viaticum', a sure guide without which nobody should set foot in the Danubian lands. He also said that as a political geographer he made no claim to be a pioneer for there were several masters of the subject already.   However, their treatment was static, concerned with the internal politics of states in relation to their environmental circumstances. To Ancel himself the political situation was never static but always dynamic; 'c'est la vie!' he wrote. Ancel's judgement is hardly fair to André Siegfried, who had travelled widely, given courses of interesting lectures, published a good deal and shown a wide understanding of the international aspects of political geography.

Some of the leading French geographers received the new recruit with coolness, despite his passionate advocacy of geography.   There was a hint of conde-scension in the letter sent by André Siegfried that the *Frontières* showed its author to possess 'undoubted competence' and perhaps also in the additional comment that 'there are a few observations that add nothing to your fine work' (in 1938). Vigorous opposition came from Jules Sion in a review in the *Annales de Géographie*, which famous journal incidentally published only two articles by Ancel (in 1925 and 1927) and no obituary. Sion recognized the hard work of Ancel but eagerly seized on mistakes, such as the statement that the railway workers in Salonica lived in rooms only 0.80m high instead of 1.80m. Apparently, said Sion, Ancel was confusing people with their poultry: such lapses were embarrassing. The book might prove useful to people interested in such problems, he observed, but it included comments on financial organization and public health which were not normally studied by geographers. (At the time Ancel was interested in paludinism.) Sion also said that insufficient treatment was given to the deltaic morphology of the area around Salonica. He went further and asked if Ancel had 'the geographical spirit' and so differed in his assessment from that of Lucien Febvre in 1945.

There was perhaps some element of jealousy engendered by the elegant production of Ancel's work as he was able to use his private means in publication. However in 1952 Jean Gottmannn made some interesting comments on Ancel's work. He found the use of the term 'isobar' for frontiers unhappy for the idea was that these isobars were in a constant state of deformation. Frontiers could not be in such a state and perhaps therefore it would seem that Ancel had to some extent absorbed some of the outlook of the geopoliticians and associated geographers. In the book on *Frontières*, Gottmann thought, the statement that the geographer knew no natural, or even linear or historical frontiers was a dangerous heresy.

In fact Ancel was never a disciple of the German geopolitical school but rather its firm opponent. He objected to the dramatic -- even sensational -- presentation of maps and atlases. The German geopoliticians relied from 1920 to 1940 on the 'suggestive map', to use the expression of Karl Haushofer himself in 1928. Ancel rejected such methods of propaganda for his maps were designed only for reference purposes and to elucidate his text. But from the Germans he acquired some ideas of substance which he proposed to use for the welfare of France.

Ancel viewed the world primarily as it affected France. Asked at Compiègne about the 'Jewish nation' he said that there was no Jewish nation. This is a typical French reaction. He was interested in the Bulgarians, whom he thought of as the Prussians of the Balkans while the Serbs and the Romanians were the French of the East. In 1922 he thought that he could see the first indications of a European society. This was Western civilization 'or perhaps it would be better to say French civilization' he added in 1928. In 1936 he looked to Geneva not as 'capital of the New Europe' but 'of the French school'. In August 1939 he spoke with conviction of his life: 'I have always respected people...the armed vigil of the present time does not prevent us from longing for a Europe in which the greatness of any one nation will not inhibit the liberty of others'.

3. INFLUENCE AND SPREAD OF IDEAS

In peacetime Ancel showed some of the same qualities that had made him successful as a soldier, for he saw that there was a marked conflict of view in political and scientific thinking and prepared his defences accordingly. From 1924 he taught economic geography at the École des Hautes Études Commerciales, and in 1926 he became a member of the selection board for candidates seeking admission to the military academy of Saint Cyr. His move to a chair at the Institut des Hautes Études Internationales of the University of Paris in 1926 was the beginning of the most fruitful period of his life. His influence was spread partly through the journal *L'Information*, of which he was the diplomatic correspondent from 1924 to 1929. His main ideas were expressed in this journal and in 1930 he published his *Histoire contemporaine (1848-1930)* for the upper forms of lycées. Several of his works were published by firms specializing in school manuals but he also wrote more substantial works such as the *Peuples et nations des Balkans* of 1926 with its supplementary study on Macedonia of 1930 and his famous *Manuel géographique de politique Européenne*, of which the final volume, on Germany, was published posthumously in 1945.

Ancel's book of 1936, *Affaires étrangères*, consisted of five chapters, the first an introduction and the last a conclusion. His second chapter was on 'La Menace', Germany; his third on 'L'Enigme', Italy, and the fourth on 'La Parade', meaning his hope of a new Europe united around France, opposed to the aggression of Germany and the opportunist policies of Italy. In his book he included an analysis of *Mein Kampf* and said that Hitler was 'a mystic, apostle and prophet' who could 'galvanize, discipline and regiment crowds'. In 1943 Robert Urban, a publicist of the Wilhelmstrasse, in his book on the underground press published in Prague, said that Ancel was a warmonger (comparable to Winston Churchill, Anthony Eden and other famous British figures of the time). Urban made much of the discovery in 'secret files' that Ancel had been paid money by the Czechs, though in fact the amounts were modest payments for Ancel's articles published in Czechoslovakia. All this, however, was more indicative of the political propaganda of the war years than of the academic search for truth.

German geopoliticians could not ignore Ancel, the author of *Géopolitique* and *Géographie des frontières*. His main opponent was Haushofer, also the author of a book on frontiers *(Grenzen)* and director of the *Zeitschrift für Geopolitik*. As early as 1932 Haushofer spoke of Ancel as a collaborator with Demangeon in the French struggle against German geopolitical expansion. Four years later, Haushofer gave his colleague Ludwig Neser the opportunity of showing admiration for the ability Ancel possessed of making a plausible thesis out of a bad case. Three years later, in 1939, Neser published a résumé of Ancel's *Géopolitique* but critical comment was to appear later. Haushofer himself wrote on Ancel's *Frontières* in 1940, with considerable respect for his material but with comments also on his errors and inconsistencies: he approved, however, of Ancel's view that frontiers should not be regarded as permanent but could be changed from time to time. (This view was held by liberally minded people in many countries during the 1930s but as noted earlier not by Jean Gottmann.)

Further attack came from Germans less well known than Haushofer, such as Franz Briel, who in 1942 spoke of the inadequacy of the 'human geography' on which Ancel's work was based and in the following year of the 'slim volume' he had written on geopolitics, 'a polemic work, entirely fragmentary'. By then Ancel was in his last years of life. His own assessment of German work was given in *Confins occidentaux*, the first part of his *Manuel géographique de politique européènne* published in 1940, where he acknowledged the scholarship, the kindly reception of his work and the liberalism of German editors, geographers and geopoliticans. His disagreement with Haushofer was marked by dignity and respect on both sides and it may be that this saved him from the 'camp of slow death' at Compiègne. As it happened, he survived for only a short time and he shared the tragic destiny of Marc Bloch (1886-1944).

Bibliography and Sources

1. REFERENCES AND SOURCES ON JACQUES ANCEL
Apparently no obituaries appeared but valuable material was found in the Mission des Archives de l'Académie de Paris, the Institut des Hautes Études Internationales, and the Centre de Documentation Juive Contemporaine. This study owes much to the staffs of these organizations and also to the author's prefaces together with the critical reviews of his work.
These include:

Sion, Jules, 'La Macédoine, d'après le livre de M. Jacques Ancel', *Ann. Géogr.*, vol 41 (1932), 305-9

Haushofer, Karl, 'Geopolitik in Abwehr und auf der Wacht', *Z. für Geopolitik*, vol 7 (1932), 594-601

Biographical note (unsigned but by Ancel) with a list of his main publications in *Recueil des Cours de l'Académie de Droit International de La Haye*, 1936, 205-6

Neser, Ludwig, 'Géographie allemande ou géographie française', *Z. für Geopolitik*, vol 13 (1936), 267 (refers to an adulatory review of Ancel's book *Géopolitique* in the Belgian journal *Le Flambeau*, March 1936)

Neser, Ludwig, 'Die französische geographische Schule und die Geopolitik' (translation of an article written by Ancel on the French school of geography and geopolitics), *Z. für Geopolitik*, vol 16 (1939), 640-56

Haushofer, Karl, 'Strebepfailer zür Geopolitik', *Z. für Geopolitik*, vol 17 (1940), 149-51

Briel, Franz, '"Geopolitik" und "géographie humaine"', *Z. für Geopolitik*, vol 19 (1942), 295-6 (also in the Belgian review *Bull. de l'Ouest*, May 1942)

Briel, Franz, 'Das geographische Denken in Frankreich' ('Geographical ideas in France'), *Z. für Geopolitik*, vol 20 (1943), 352-5

Urban, Rudolf, *Demokratenpresse im Lichte Pragar Geheimakten* (The democratic press of the underground movement), Prague, (1943), 328p.

Bernard, Jean Jacques, *Compiègne 1941-1942, le camp de la mort lente*, Paris (1944), 250p.

Febvre, L., 'Slaves et Germains' *Ann. Hist. Soc.*, vol 7 (1945), 147-8

Gottmann, J., *La politique des états et leur géographie*, Paris (1952), 228p.

Bernard, J.J., *Mon père, Tristan Bernard*, Paris (1955), 272p.

Coston, H., *Les causes cachées de la 2e Guerre mondiale*, Paris (1975), 224p.

2. WORKS BY JACQUES ANCEL
1919 *L'unité de la politique bulgare (1870-1919)*, Paris, 80p.

1921 *Les travaux et les jours de l'Armée d'Orient (1915-1918)*, Paris, 234p.

1922 'La géographie de l'histoire', *La Géogr.*, vol 37, 439-516

1923 *Manuel historique de la question d'Orient (1792-1923)*, Paris, 336p.

1925 'Les migrations des peuples dans la Grèce actuelle', *Ann. Géogr.*, vol 34, 277-80

1926 *Peuples et nations des Balkans*, Paris, 220p. (*Ann. Géogr.*, vol 36 (1927), 74-6). See also 1930, *La Macédoine, son évolution contemporaine*.

1928 *Les Balkans face à l'Italie*, Paris, 130p.

1929 (ed H. Hauser) *Histoire diplomatique de l'Europe (1871-1914)* 2 vols, 476p. and 390p. Ancel's contributions are in vol 2

1930 *Histoire contemporaine (1848-1930)*, Paris, 604p. Reissued in 1934 to cover 1815-1930, cxviii + 604p.

1930 *La Macédoine, son évolution contemporaine*, Paris, 352p.

1931 'Géographie politique de la Baltique', in Pages, G. *et al.*, *La Pologne et la Baltique*, Paris, 84-98

1932 'Géographie politique de la Haute Silésie', in Eisenmann, L. *et al.*, *La Silésie Polonaise*, Paris, 37-54

1933 'Géographie politique de la Prusse Orientale', in Ancel, J. *et al.*, *La Pologne et la Prusse Orientale*, Paris, 21-43

1934 *Les frontières slaviques*, Dijon, 40p. (also in *Monde Slave*, April-May 1934)

1935 *Les frontières roumaines*, Bucharest, 70p. (also in *Rev. Hist. du Sud-Est Eur.*, January-March 1935)

1936 *Affaires étrangères, aide-mémoire de la politique française (1789-1936)*, Paris, 128p.

1936 *Géopolitique*, Paris, 120p. (an introduction to the volume mentioned below)

1936 *Manuel géographique de politique européènne*, vol 1, L'Europe centrale, Paris, 404p.

1938 *Géographie des frontières*, Paris, 210p.

1940 *Manuel géographique de politique européènne*, vol 2/1, L'Europe germanique et ses bornes. Les confins occidentaux, Paris, 245p.

1945 *ibid.*, vol 2/2, L'Allemagne, Paris, 270p.

1945 *Slaves et Germains*, Paris 224p.
The last two works are posthumous.

Robert Specklin is a research worker at the Centre de Géographie Appliquée, Université Louis Pasteur, Strasbourg, France. Translated by T.W. Freeman.

CHRONOLOGICAL TABLE: JACQUES ANCEL

DATES	LIFE AND CAREER	ACTIVITIES, TRAVEL, FIELDWORK	PUBLICATIONS	CONTEMPORARY EVENTS AND PUBLICATIONS
1882	Born 22 July at Parmain			
1897				Ratzel, *Politische Geographie*
1905	Teacher at Vannes			
1906	Teacher at Péronne			
1908	Agrégé			
1909	Teacher at Chaptal			
1914	Called up in Lorraine			Outbreak of war
1916		Went to the eastern front		
1919	Returned to Chaptal		*Politique bulgare*	
1921			*Travaux et jours*	Brunhes and Vallaux, **Géographie de l'histoire**
1922			'Géographie de l'histoire'	
1923			*Question d'Orient*	
1924	Lecturer at École des Hautes Études Commerciales	Worked for *L'Information* (to 1929)		*Zeitschrift für Geopolitik* founded
1925				Maull, *Politische Geographie*
1926	Left Chaptal to teach at the Institut des Hautes Études Internationales, Univ. of Paris			
1927	Began association with the Carnegie Centre Européen de la Dotation	Toured Eastern Europe (to 1929)		Haushofer, *Grenzen*
1928			*Les Balkans face à l'Italie*	Hennig, *Geopolitik*
1929				
1929			*Histoire diplomatique de l'Europe*	
1930	Doctorate		*La Macédoine*	
1931		Further travels in Eastern Europe (to 1933)	'La Baltique'	
1932		Corresponding member of the Romanian Academy of Science	'La Haute Silésie'	
1933		Corresponding member of the Belgrade Geographical Society	'La Prusse Orientale'	Hitler régime in Germany
1934		Secretary-General, Comité d'Études de l'Europe Centrale	*Les frontières slaviques*	

DATES	LIFE AND CAREER	ACTIVITIES, TRAVEL, FIELDWORK	PUBLICATIONS	CONTEMPORARY EVENTS AND PUBLICATIONS
1935			*Les frontières roumaines*	
1936			*Géopolitique* and *Affaires étrangères*	
1938		Visited Salonica in August	*Frontières*	
1939				Germany attacks Poland
1940			*Confins occidentaux*	
1941	Interned at Compiègne			
1942	Released			
1943	Died December			

Philippe Arbos
1882-1956

MAX DERRUAU

1. EDUCATION, LIFE AND WORK

Philippe Arbos was born in the French village of
Mosset in the eastern Pyrenees situated 600 metres
above sea-level, to which he returned regularly
throughout his life. From his father, a school
teacher, came his republican, socialistic but non-
Marxist, and secular ideas. Philippe Arbos kept his
Catalan accent throughout his life but was opposed to
regionalism as an ideology. He went as a pupil to
the primary school where his father taught, and later
to lycées at Perpignan and -- to prepare himself for
entry to the École Normale Supérieure -- at Toulouse
and Paris (Louis-le-Grand). He became a scholarship
holder for two years (1902-4) at the Sorbonne and,
having achieved success in the competitive entry, in
1904 he entered the École Normale Supérieure. His
liberal ideas were strengthened by the Dreyfus case
but though he was involved in politics in his home
area he never became a candidate for election.

He learnt German, which he read very well, at the
lycée and English at the École Normale but though he
could read English fluently he never spoke it. As a
student he followed some courses given by Vidal de la
Blache, but more by Lucien Gallois. Of greater
significance in his development, however, was his
friendship with Raoul Blanchard, then a young university teacher at Grenoble and shortly to become head of
the Geography Department, one of the two French geography 'schools'. Blanchard was a member of the
board of examiners for the *agrégation*, so they met
first as examiner and candidate. Though they became
devoted friends later, Blanchard remained the *maître*.
Arbos attained the *agrégation* in history and geography

in 1907 and was appointed to the lycée in Toulon but he
feared that he might vegetate in a Mediterranean town
and so, like Blanchard, he settled in Grenoble in
1909. He taught for ten years at the lycée in
Grenoble, where he met his first wife, who taught
literature in a girls' school. The only child of the
marriage, Lucienne (Mme Gérard Bloch), also became a
teacher of literature.

Under the direction of Raoul Blanchard, Arbos
chose as his thesis subject 'La vie pastorale dans les
Alpes françaises' and in his Alpine cloak he walked
the mountains collecting data. Having contracted
pleurisy on the eve of the 1914-18 war he was dis-
pensed from military obligations but when cured he was
drafted into an auxiliary service at Grenoble from
1915 to 1917. He made a point of proceeding slowly
with his thesis so that he would not benefit from any
promotion denied to his contemporaries who were
fighting. The presentation of his work was delayed
to 1922, although the documentary research was virtually complete by 1914.

Meanwhile, in 1919, he was appointed to the University of Clermont-Ferrand where the Geography Institute had been staffed only intermittently by teachers,
of whom the last, Léon Boutry, had died during the
war. Arbos stayed at Clermont-Ferrand for 33 years,
during which time he became enthralled by the geography of the Massif Central, on which he published several regional articles. He also wrote three articles
on Clermont-Ferrand from 1925 to 1929, which marked an
advance in urban geography at the time, as well as a
book on the Auvergne in 1932. He retained his status
as the French expert on pastoral life. His main

influence on others was acquired through his writing, but he had a gift for friendship with his colleagues and became deputy to the Dean of the Faculty in 1937.

He attended the International Geographical Congresses at Paris in 1931 (during which he directed the Auvergne excursion), at Warsaw in 1934 and Amsterdam in 1938. His other travels included two visits to North Africa, the teaching of a summer course at Middleburg, Vermont, U.S.A., in 1927, a tour of central Europe with Max Sorre in 1934, and a semester's work at Rio de Janeiro in 1937 which resulted in his 'Petropolis' article of 1938.

The death of his first wife in 1936 and the war of 1939 to 1945 saddened Arbos considerably. The defeat of 1940 and the Vichy régime (during which his daughter was arrested), the German occupation and the interferences with human rights after the liberation, all accentuated his natural pessimism. His second marriage, in 1946, gave him a home once more after ten solitary years. In 1952 he retired to Andancette, Drôme, where his wife was a pharmacist, though for a further year he gave courses in the Faculté des Lettres at Clermont-Ferrand. The 'jubilee' which was celebrated in 1953 by the publication of two volumes in his honour, was in honour of his seventieth birthday (1952) and of the anniversary of his doctorate (1922). Unfortunately he was already ill and his death came in October 1956 at Andancette.

2. *SCIENTIFIC IDEAS AND GEOGRAPHICAL THOUGHT*
Though known mainly as a specialist on pastoral life throughout the world but especially in the Alps, Philippe Arbos was also interested both in urban geography, on which he published papers dealing with Clermont-Ferrand, Skoplje and Persepolis, and in regional geography. Geomorphology was not a main interest, though he followed closely all the work done on the glaciation of the Massif Central.

He agreed with the possibilist views of Vidal de la Blache and with the humanist outlook of French geographers at the end of the nineteenth century. The general economy of states was of only minor interest to him: he cared much more for the regional aspects, especially of the *pays*. Basic to all his work was the study of the adaptation of man to the environment and in this the physical aspects must be considered and described, but not subject to morphological analysis. Far more significant was historical evolution, from an old agricultural economy existing before the Agricultural Revolution to modern farming.

He found that Marc Bloch and Roger Dion had much of interest to say on agricultural landscapes and he used their writings in preparing his lectures. To some extent also he followed their research methods, notably their use of cadastral map sources. He did this to a minor degree in his thesis of 1922 but to an increasing extent later, perhaps under the influence of his pupil Lucien Gachon. But he feared that the researches on agrarian history could result in a form of regionalism that might become almost as dangerous politically as racialism, possibly resulting in the fission of states, especially of France. In general he was more interested in the practice of arable and pastoral farming than in the field pattern.

In his work on demographic problems he held the view that the recognition of regions by distributional mapping on a basis of administrative units was not geographically sound. Always in his work he was well aware of contingency: nature offered possibilities but the use of them depended on human choice. The articles on urban geography, notably on Clermont-Ferrand, show this clearly. In Clermont-Ferrand the pneumatic tyre industry had been established and expanded through chance circumstances, to use his own phrase a *'cascade de hasards'*.

Firmly opposed to any use of geography for propaganda purposes, such as that associated with pan-Germanism, Arbos found himself confronted by many new developments, such as the use of laboratory work in physical geography and the political commitment assumed by a number of geographers. Though he had a keen appreciation of the difficulties of people everywhere, he remained convinced that political salvation could be achieved only through the liberalism of the republican and 'radical' (in the French sense) tradition of which he was an admirer from his childhood.

3. *INFLUENCE AND SPREAD OF IDEAS*
Philippe Arbos achieved a position of great respect and influence at the University of Clermont-Ferrand through his work there from 1919 to 1952, indeed to 1953. There he taught successive generations of students who became valued teachers in various forms of secondary education. Beyond the Faculty he was an active member of the learned society known as 'Les Amis de l'Université' and he also gave a number of lecture courses open to the public on the Massif Central and on other regions of the world. His teaching was given with particular clarity and his work in the classroom was supplemented by numerous excursions, some of them arranged with the help of historians. A man of friendly temperament, he welcomed any student who cared to call at his home during the early part of the afternoon.

His first student was Lucien Gachon, a novelist and schoolmaster who later taught in the Faculté de Lettres at Clermont and in time became a professor in the universities of Clermont and Besançon and a member of the Academy of Agriculture. Under the direction of Arbos, Gachon presented his thesis on *Les Limagnes du Sud et leurs bordures montagneuses* in 1939. Arbos directed other theses, including those of Alfred Durand on *La vie rurale dans les massifs volcaniques des Dores, du Cézallier, du Cantal et de l'Aubrac* in 1946 and of Max Derruau on *La Grande Limagne auvergnate et bourbonnaise* in 1949. Earlier, by chance circumstances, he assisted Jean Fischer to prepare his thesis on *L'Adour et ses affluents*, presented in 1929. He also followed sympathetically the normal trials of an author endured by his former student André Fel in writing *Les hautes terres du Massif Central, tradition paysanne et économie agricole*, though this thesis was not submitted and presented until after his death. Among his other pupils, mention should be made of Henri Onde, who became a professor at the University of Lausanne, and Pierre Coutin, from farming stock in the Limagne, who became, after a period at the Ministry of Agriculture, a professor at the École Pratique des Hautes Études.

Philippe Arbos also influenced students at the École Normale Supérieure, Saint-Cloud, where he gave courses, and he was known and respected by a number of Brazilian geographers.

A man of natural modesty, Arbos never tried to impose any doctrine on others and his influence was somewhat diffuse, especially as he always tried to encourage a critical approach to methods of work among his students and others. In general he made a firm contribution to possibilism in geography, and so was in the tradition of Vidal de la Blache and Albert Demangeon, followed later by Roger Dion and Marc Bloch.

Bibliography and Sources

1. REFERENCES ON PHILIPPE ARBOS
Blanchard, R., 'Philippe Arbos', *Rev. Géogr. Lyon*, vol 23 (1957), 57-8
Gachon, L., 'Philippe Arbos', *Rev. Géogr. Alp.*, vol 45 (1957), 5-7
Sorre, M., Philippe Arbos, *Ann. Géogr.*, vol 66 (1957), 182-3

2. WORKS BY PHILIPPE ARBOS
The published works of Arbos reveal his two major specialisms: pastoral and Alpine life, and the Massif Central. Most of the publications on the first of these themes are contemporary with his thesis, mainly before 1923. The second group dates from the time when he was at the University of Clermont-Ferrand, and particularly from 1922-37. They deal mainly with human and regional geography, though the latter interest led on occasion to researches into economic geography and geomorphology. There were also some publications on the Massif Central intended for the general public. Urban geography is represented by five articles, three of which deal with Clermont-Ferrand.

His homeland, the eastern Pyrenees, inspired one of Arbos's first articles. He was also a contributor of notes to a number of journals on a wide range of topics, even including coal in the U.S.A. He translated part of vol 3 of the famous textbook by A. Penck and E. Bruckner, *Die Alpen im Eiszeitalter (The Alps in the Ice Age)*.

Among the journals for which he wrote were the *Travaux de l'Institut de Géographie Alpine*, which in 1919 became the *Revue de Géographie Alpine*, and the *Annales de Géographie*. He contributed articles on occasion to local reviews, such as the *Almanach de Brioude* and also to journals published outside France, for example the *Geographical Review* and the *Bulletin de la Société Belge d'Études Géographiques*. For a fuller list of his publications readers may consult the volumes of the *Bibliographie internationale*.

Several of his papers were collected together in the jubilee volume of 1953, *Mélanges géographiques offert a Ph. Arbos*, Publications de la Faculté des Lettres de l'Université de Clermont-Ferrand, fasc. 7 et 8, Institut de Géographie III et IV, 220 and 304p., 1953. These volumes also include a number of articles by his pupils and friends but among those by Arbos are the three articles on Clermont-Ferrand and an article on the Margeride which appeared in the *Almanach de Brioude*.

1909 'Les glaciations des Alpes du Sud', translation of part of vol 3 of Penck and Bruckner, *op. cit.*, *Ann. Univ. Grenoble*, vol 19, 118-67
1910 'La Plaine du Roussillon, *Ann. Géogr.*, vol 19, 150-68
1919 'La légende du Devoluy', *Trav. Inst. Géogr. Alp.*, vol 7, 389-95
1922 *La vie pastorale dans les Alpes françaises: étude de géographie humaine* (thesis), Faculté de Lettres Grenoble, Paris, 718 p.
1922 'Questionnaire pour l'étude de la vie pastorale en montagne', *Rev. Géogr. Alp.*, vol 10, 489-94
1922 'Le glaciaire dans le Massif Central', *Rev. Géogr. Alp.*, vol 10, 481-7. See also *ibid.*, 'Morphologie glaciaire aux environs de Bort (Corrèze)', vol 15 (1927), 343-5
1923 'The geography of pastoral life illustrated with European examples', *Géogr. Rev.*, vol 13, 559-74
1923 'L'industrie de la dentelle dans la région du Puy et son évolution actuelle', *Ann. Géogr.*, vol 32, 353-5. See also *ibid.*, vol 43 (1934), 191-3 and 'L'industrie dans le Velay du Nord-Est', *ibid.*, vol 44 (1935), 416-20
1924 'Excursion géographique (en Auvergne)', *Ann. Géogr.*, vol 33, 404-6. See also under 1933
1925 'Le Cadre géographique de Clermont-Ferrand', *Rev. Géogr. Alp.*, vol 13, 623-56
1927 'Clermont-Ferrand: les fonctions urbaines', *Rev. Géogr. Alp.*, vol 15, 375-454
1929 'Clermont-Ferrand: l'organisme urbain', *Rev. Géogr. Alp.*, vol 17, 289-328

NOTE - These three papers were published in a volume of 160 pages at Clermont-Ferrand in 1930 and reproduced in a shortened form in the *Mélanges Arbos* of 1953, listed above.

1926 'Le massif du Cézallier; étude de géographie humaine dans la montagne d'Auvergne', *Rev. Géogr. Alp.*, vol 14, 573-99
1927 (in collaboration with L. Gachon) *Nouvelle Géographie du Puy-de-Dôme*, Clermont-Ferrand, 36 p.
1928 (in collaboration with André Meynier), 'La Châtaigneraie cantalienne', *Bull. Assoc. Géogr. Fr.*, April-May, 44-7
1929 'Questionnaire d'enquête géographique', *Rev. d'Auvergne*, vol 43, 11-18
1929 'Quelques types de maisons rurales en Limagne', *Bull. Assoc. Géogr. Fr.*, June, 63-4
1930 'Les migrations internationales aux XIXe et XXe siècle', *Ann. Géogr.*, vol 39, 84-8
1930 'Skoplje', in 'Excursion interuniversitaire en Yougoslavie', *Ann. Géogr.*, vol 39, 324-8
1931 'La Margeride', *Almanach de Brioude et de son arrondissement*, vol 13, 86-101
1932 *L'Auvergne*, Paris, 224 p.
1932 'L'Artense', *Rev. Géogr. Alp.*, vol 20, 677-700
1933 'L'Auvergne, les genres de vie d'une province française', *Bull. Soc. Belge d'Étud. Géogr.*, vol 3, 57-79
1933 'La XXIVe excursion géographique interuniversitaire', *Ann. Géogr.*, vol 42, 529-31

1933 (in collaboration) *Les populations rurales du Puy-de-Dôme*, Mémoires de l'Académie des Sciences, Belles-Lettres et Arts de Clermont-Ferrand, tome 93 de la Collection Annales et Mémoires, 439 p.

1934 'Un rapport du préfet Ramond sur l'émigration saisonnière dans le département du Puy-de-Dôme', *Rev. d'Auvergne*, vol 48, 89-102

1937 'La houille blanche dans le Massif Central', *Bull. Soc. Belge d'Étud. Géogr.*, vol 7, 66-85

1938 'Petropolis, esquisse de geographie urbaine', *Rev. Géogr. Alp.*, vol 26, 477-530

1950 'Le declin des migrations saisonnières dans une vallée des Pyrénées méditerranéennes', in *Livre jubilaire offert à M. Zimmermann* , Lyon, 151-4

Max Derruau is Professor of Geography at the University of Clermont-Ferrand, France.   Translated by T.W. Freeman.

CHRONOLOGICAL TABLE: PHILIPPE ARBOS

DATES	LIFE AND CAREER	ACTIVITIES, TRAVEL, FIELDWORK	PUBLICATIONS	CONTEMPORARY EVENTS
1882	Born 30 July at Mosset, eastern Pyrenees			
1893-9	Pupil at the lycée, Perpignan			
1899-1902	Studied at Toulouse and Paris (Louis-le-Grand)			
1902-4	Student with scholarship at the Sorbonne			
1904-7	Attended École Normale Supérieure, rue d'Ulm, Paris			
1907	*Agrégation* in history and geography; met Raoul Blanchard at the examination			
1907-9	Taught at the lycée in Toulon			
1909	Appointed to the lycée at Grenoble	Began work on his thesis, 'La vie pastorale dans les Alpes françaises'		
1910			Published first article, 'La plaine du Rousillon', *Ann. Géogr.*	
1914				First World War
1915	Served in auxiliary forces but still resident in Grenoble (to 1917)			
1918				End of war
1919	Appointed to Univ. of Clermont-Ferrand			
1922		Presented thesis	Publication of thesis in Grenoble and Paris	
1925			The first of three articles on Clermont-Ferrand (others 1927 and 1929) appeared	
1927	Taught in a summer course at Middleburg, Vermont, U.S.A.			
1931		During I.G.U. Congress organized and led an excursion in the Auvergne		International Geographical Congress, Paris

DATES	LIFE AND CAREER	ACTIVITIES, TRAVEL, FIELDWORK	PUBLICATIONS	CONTEMPORARY EVENTS
1932			First book on the Auvergne	
1934		At I.G.U. Congress; toured Central Europe with Max Sorre		International Geographical Congress, Warsaw
1936	Death of Mme Arbos			
1937	Deputy to Dean of the Faculty	Gave a course for one semester at Rio de Janeiro		
1938		At I.G.U. Congress	'Petropolis' article	International Geographical Congress, Amsterdam
1939				Second World War
1940				Fall of France and German occupation
1945				End of war
1946	Second marriage			
1952	Retired from chair at Univ. of Clermont-Ferrand and moved to Andancette, Drôme	Some teaching given at Clermont-Ferrand in his first year of retirement		
1953	Celebration of his jubilee		*Mélanges Arbos*, 2 vols, to celebrate his career	
1956	Died in October at Andancette			

Wallace Walter Atwood

1872–1949

WILLIAM A. KOELSCH

Wallace Walter Atwood, geologist, physiographer, and university president, was in his own time a dominant figure in American geography; a contemporary memorialist accurately described his life's work as 'a monadnock in the landscape of his generation'. Even during his lifetime, however, the conceptual framework of his special science of geomorphology was shifting and the 'new geography' he taught was already becoming obsolescent. Yet Atwood probably brought geography to more people than any other American, and in this respect he was a significant figure in the history of geographical education.

1. EDUCATION, LIFE AND WORK

Wallace Atwood, born on 1 October 1872 in Chicago, Illinois, was the son of the owner of a planing mill and the descendant of an old Massachusetts family. After graduating from the city's West Division High School he matriculated at the new University of Chicago in December 1892. The most important intellectual influence on him there was the geographer-geologist Rollin D. Salisbury, under whom Atwood took a summer field course in the Devil's Lake region near Baraboo, Wisconsin. This experience was the key element in his early commitment to a career of systematic field study as a geographer-geologist.

During his graduate years in geology at Chicago, Atwood held a number of part-time teaching positions in pre-collegiate and teacher-training schools associated with the University. Here he worked under the direction of two noted educational reformers, Colonel Francis W. Parker and John Dewey, whose ideas on child development greatly influenced Atwood. His marriage in 1900 to a fellow teacher in the University's laboratory school system brought him an essential partner and critic in his work of improving pre-college teaching, as well as a companion on his world travels and field studies. Atwood was named Instructor in Physiography and General Geology in the University in 1901, and held the rank of Associate Professor at the time of his departure for Harvard University a dozen years later.

Atwood's first publication, written jointly with his mentor Salisbury, dealt with the geography and geology of the Devil's Lake region of Wisconsin, where Atwood himself soon began to conduct field courses. He also served as Salisbury's assistant on the New Jersey Geological Survey and on the Wisconsin Natural History Survey, and later served with the Illinois State Geological Survey. His appointment in 1901 to the U.S. Geological Survey led him to three seasons in Alaska and then into the Rocky Mountains. In 1909 he was commissioned by the Survey to make a lengthy study of the San Juan Mountains of Colorado, which resulted in a large number of scientific papers and a major monograph.

Atwood was appointed Professor of Physiography at Harvard University in 1913, carrying on a tradition of landform studies initiated by William Morris Davis. Atwood revitalized the teaching of physical geography at Harvard, developing advanced courses for increasing numbers of enthusiastic undergraduates, using local field excursions and teaching systematic advanced field courses, including summer field experiences in the mountains of the West. Throughout his teaching career his emphasis on the primacy of field studies was to be a significant influence on student thinking 'in the field'. During his Cambridge years Atwood also contracted with the Boston firm of Ginn and Company to revise a series of school geographies begun by Alexis E. Frye. He developed a close relationship with Charles H. Thurber, a Ginn editor who was also chairman of the Clark University Board of Trustees. Early in 1920 Atwood accepted the Presidency of Clark University, in Worcester, Massachusetts, with a specific mandate to develop graduate and undergraduate programmes in geography and related fields under his personal direction. Atwood did not retire until 1946, at the age of 73.

2. SCIENTIFIC IDEAS AND GEOGRAPHICAL THOUGHT

Atwood's principal contributions to science were his geomorphological studies at the Rocky Mountains. His long association with the U.S. Geological Survey enabled him to spend most of his summers in that part of the world, a practice continued through a 17,000 mile field survey in the summer of 1948 when he was approaching his 76th birthday. His major scientific publication was a substantial monograph on the San Juan Mountains (with K.F. Mather), completed in 1925 though not published until 1932; one modern writer has called it 'a classic of its kind'. In the 1920s and 1930s Atwood moved outward from the San Juans to other sections of the southern Rockies and northward into Montana, publishing a number of technical papers in collaboration with his son, Wallace Jr. In 1938 he presented a 'working hypothesis' on the physiographic history of the entire region, but his projected 'final report' on Rocky Mountain Quaternary geomorphology was never completed. His general findings and approach, however, are indicated in his textbook on the physiographic provinces of North America incorporating his university lectures, and his work on the Rockies written for the lay reader.

As a physiographer, Atwood's primary scientific purpose was to reconstruct the physical history of the land on terms acceptable to geologists through the use of meticulous field observation and analysis. His methods were, as he put it in his San Juan study, 'fundamentally geologic': field reconnaissance, followed by detailed field examination (including mountain climbing); both intensive and generalized mapping of surface forms ('the sculpture of which has made the geography'); genetic classification of the features observed; analysis of these landforms in relation to other areas and the existing literature; explanation of key features and any anomalies or special problems; and detailed reporting of the results in the form of oral presentations at scientific meetings, abstracts, articles and survey monographs.

In his monographs, Atwood would describe the scenery and try to explain it in terms of 'the sequence of events in the physiographic history of the region'. His discussion was lucid and well organized and characteristically emphasized the visible forms rather than the processes. He was an exceptionally competent field observer, but he was not an innovator in concept, method or vocabulary, and he only rarely attempted to generalize his observations into what he called 'working hypotheses'. Yet his scientific work

on mountain glaciation was recognized in a 1941 survey of American attainments in physiography as 'work comparable to any in the world'.

Atwood's ideas on geography are an extension of his earliest field experiences. 'Out-of-doors' was a laboratory of mountains, climate, soils, natural resources and people in interaction. He saw the study of landforms and landform regions as the scientific foundation of human geography, yet he moved further than his mentor Salisbury in expounding and elaborating the influence of physical environmental characteristics on human activities in specific regions. This focus, in keeping with the environmentalist stress of early twentieth century American geography, enabled Atwood to exercise a degree of leadership in geography during the 1920s, a position reinforced by the institutionalization of his ideas in the curriculum of his newly created major graduate training programme, and by the increasing separation of geography from geology, itself being transformed by new geophysical data and new theories of matter. Ironically, the geographical ideas he expounded at the height of his influence were already under siege from his younger contemporaries during the 1920s and 1930s.

Geography, as Atwood saw it, was a science of relationships linking man to the natural environment, expressed in terms of the 'great human dramas' of men responding or adapting to natural advantages, and partially transcending natural disadvantages through the development of interdependent networks of communication and trade between distinct 'natural regions'. In Atwood's world-view, nature existed apart from the destinies of man and controlled them. 'Man does not conquer nature,' he wrote. 'He may discover the laws of nature and accomplish better and better adjustments to natural conditions on this earth.' Atwood's perspective was shaped by an oft-repeated master image of the earth as a patchwork of component natural regions, or 'natural geographic provinces', which were 'the stage setting of real human dramas'. Yet for Atwood the physical setting not only influenced but to some extent also directed human activity and although the informed members in any community might have a greater understanding of their opportunities in their physical environment, they had nevertheless to act in accordance with the qualities of the physical environment, expressed through national economic development and national character. Their choice of action was therefore limited, except in detail, though there was always the possibility of migration to a different natural region.

Atwood's view of geography was clearly conditioned by his interest in preparing Americans for an era which would demand of them certain responsibilities towards their own and other countries of the world. The task of the geographer was to sensitize schoolchildren, teachers, college and university students, government officials and businessmen to the importance of geographical conditions in explaining the economic and social setting of human activity. For, Atwood asserted, 'Nature has determined through variety in soils, in landscapes, in climate, and in peoples, the interdependence of one part of the earth upon another and of one people upon the activities of another.' In all of his major public addresses from 1919 onwards Atwood stressed the importance of geographical understanding in 'the establishment of good will among the peoples of the earth', and therefore in the peaceful solution of world problems.

A contemporary problem which appealed to Atwood was the protection and wise use of natural resources, and in his view geography and geographers could make a significant contribution towards solving it. Early research for the U.S. Geological Survey in the economic geology of parts of Alaska and the Colorado Plateau had given him insights into mineral resources. During his studies in the Colorado Rockies, he made observations of the old mining towns, as well as the management of grazing lands, timber resources and water. He developed a deep commitment to the National Park system from its earliest days, and he was a tireless advocate of conservation and wise use of the resources a bountiful Nature had provided for the benefit of man. Indeed, Atwood established the internationally circulated journal *Economic Geography* in 1925 in part to appeal to 'all who are engaged in intelligent utilization of the world's resources'.

3. INFLUENCE AND SPREAD OF IDEAS

Atwood's major contributions to geography were institutional and pedagogic rather than to the intellectual development of the science. The first of these was the creation, organization and staffing of the Graduate School of Geography at Clark, beginning in 1921. The second fully staffed, independent Ph.D. granting department to be established in that field in America, it quickly became a major centre for instruction in geography, placing special emphasis on fieldwork, physical geography as the foundation of all other geographies, land use studies, and geographical education. The early staff included such outstanding geographers as E.C. Semple, C.F. Brooks, C.F. Jones and, on a part-time basis, O.E. Baker, C.F. Marbut, and H.L. Shantz.

Hundreds of students were attracted there from more than a score of countries during the Atwood years, and some 70 Ph.Ds and nearly 200 M.A.s were awarded by the School while he was its Director. Their holders in many cases moved on to positions of influence in state universities and teacher-training institutions, and in government service. Atwood himself was an inspiring teacher in field and classroom, notable for his personal interest in students, for an enormously effective lecture style, and for his celebrated technique of sketching landforms on the blackboard using both hands, simultaneously illustrating structure and surface features.

Atwood's second major contribution was the advancement of geographical education at school level, primarily through a stream of widely used textbooks, with associated workbooks, study guides, wall maps and map manuals. These texts were organized on the basis of natural regions, defined as 'portion(s) of the earth's surface throughout which the geographic conditions which influence life do not differ greatly'. This organization provided an interpretive framework for the problem method of instruction, with its emphasis on understanding the significance of geographical facts rather than memorization of detail. Atwood's first text, *New Geography - Book Two*,

sold nearly four million copies in its first 25 years in print. Well before his retirement it was reported that over ten million copies of his texts had been sold, and it was estimated at the time of his death that one or more of these texts and related materials had been purchased for use in some 27,000 school districts, and that it had been used by not less than fifty million students. The simple, clear physical-political maps he designed both for text and for the classroom wall were a main support of the teaching plan. All materials were designed in a graded sequence which would match the pupil's intellectual development. The end was better citizenship through an imaginative and sympathetic understanding of the conditions determining the livelihood patterns of American and other peoples.

Atwood made a third significant contribution to geography through his skills as an evangelist for the diffusion and application of geographical knowledge. He lectured before educational, civic and environmental groups in every state, organized a New England Geography Conference in 1922, and was active in other societies and associations interested in the study and teaching of geography. As early as 1919 he became interested in the production of motion pictures for use in geographical education, and after his retirement he continued to be active in a project of the American Council on Education to produce a series of world geography teaching films. Both in Chicago and in Cambridge he was a pioneer in museum education: arranging plants and animals in natural habitat groups, developing links between museums and schools, and building the Atwood Celestial Sphere, a forerunner of modern planetaria.

As a geographer, Atwood helped to guide American youth towards an ecological awareness and an international point of view. At a time of institutional weakness of geography in the American university, he built the foundations of a centre for advanced geographical study and founded a major scholarly publication. His record of innovation in the advancement of popular geographical knowledge through untraditional teaching aids such as films, visits to museums and study tours; his enormous impact on school geography; his conservation efforts and his organizing talents earn him an honoured place in the history of American geography. He was an educationalist in the broadest sense, and through his work many became interested in the social and applied aspects of geography.

Bibliography and Sources

1. REFERENCES ON W.W. ATWOOD

There is no critical biographical study. Biographical sketches, in some cases with extensive bibliographies, are given in:

Cressey, G.B., 'Wallace W. Atwood, 1872-1949', *Ann. Assoc.Am.Geogr.*, vol 39 (1949), 296-306

Van Valkenburg, S., 'Wallace W. Atwood', *Geogr. Rev.*, vol 39 (1949), 675-7

Mather, K.F.,   'Memorial to Wallace Walter Atwood',
   Proc. Geol. Soc. Am. (1949), 106-12
Koelsch, W.A.,   'Atwood, Wallace Walter', *Dict. Am.
   Biog.* Suppl. 4 (1974), 31-2

2. PRINCIPAL WORKS OF W.W. ATWOOD

a. Geomorphology of North America
1897 (with R.D. Salisbury) 'Drift phenomena in the
   vicinity of Devil's Lake and Baraboo, Wisconsin',
   J. Geol., vol 5, 131-47
1900 (with R.D. Salisbury) 'The geography of the
   region about Devil's Lake and the Dalles of
   Wisconsin', *Wisconsin Geol. and Nat. Hist. Surv.*
   bull. no. 5, State of Wisconsin, Madison, 151p.
1903 *Glaciation of the Uinta Mountains*, Ph.D. Diss.,
   Dept. of Geology, University of Chicago
1908 (with R.D. Salisbury) 'The interpretation of
   topographic maps', *U.S. Geol. Surv.* Prof. Pap.
   no. 60, G.P.O., Washington, 84p.
1909 'Glaciation of the Uinta and Wasatch Mountains',
   U.S. Geol. Surv. Prof. Pap. no. 61, 96p.
1912 (with K.F. Mather) 'The evidence of three distinct glacial epochs in the Pleistocene history
   of the San Juan Mountains, Colorado', *J. Geol.*,
   vol 20, 385-409
1932 (with K.F. Mather) 'Physiography and Quaternary
   geology of the San Juan Mountains, Colorado',
   U.S. Geol. Surv. Prof. Pap. no. 166, 176p.
1933 'The correlation of ancient erosion surfaces',
   C.R. Congr. Interm. Géogr., Paris (1931), 588-91
1938 (with W.W. Atwood, Jr) 'Working hypothesis for
   the physiographic history of the Rocky Mountain
   region', *Geol. Soc. Am. Bull.*, vol 49, 967-80
1940 *The physiographic provinces of North America*,
   Boston, 536p.
1945 *The Rocky Mountains* ('American Mountain Series')
   New York, 324p.

b. School geographies and educational aids
1920 *New Geography - Book Two* (Frye-Atwood Geographical Series), Boston, 304p.
1921 (with H.G. Thomas) *Teaching the New Geography:
   A Manual for use with the Frye-Atwood Geographical Series*, Boston 303p.
1925 (with Harriet T.B. Atwood), *The Problem Method
   in Comparative Map Studies*, Boston, 71p.
1928 (with H.G. Thomas), *Home Life in Faraway Lands*,
   Boston, 166p.
1928 Series of ten regional-political wall maps of
   the continents and the world, Chicago
1929 (with H.G. Thomas) *The Americas*, Boston, 316p.
1930 (with H.G. Thomas) *Nations Beyond the Seas*,
   Boston, 362p.
1931 *The World at Work*, Boston, 344p.
1936 (with H.G. Thomas) *The Growth of Nations*,
   Boston, 388p.
1943 (with H.G. Thomas) *The American Nations*, Boston,
   387p.
1946 (with H.G. Thomas) *Nations Overseas*, Boston,
   392p.

c. Geographical articles, addresses and reports
1911 'A geographic study of the Mesa Verde', *Ann.
   Assoc. Am. Geogr.*, vol 1, 95-100
1919 'Geography in America', *Geogr. Rev.*, vol 7,
   36-43
1921 'The new meaning of geography in American education'(inaugural address), *Publ. Clark Univ.
   Lib.*, vol 6, 25-37, also *Sch. and Soc.*,
   vol 13, 211-18

*William A. Koelsch is Professor of Geography at
Clark University, Worcester, Massachusetts, U.S.A.*

CHRONOLOGICAL TABLE: WALLACE WALTER ATWOOD

DATES	LIFE AND CAREER	ACTIVITIES, TRAVEL, FIELDWORK	PUBLICATIONS	CONTEMPORARY EVENTS
1872	Born 1 October in Chicago, Ill.			
1892	Matriculated Uni. of Chicago			
1897	S.B.degree	New Jersey Geological Survey	First published paper: 'Drift phenomena in the vicinity of Devil's Lake' (with R.D. Salisbury)	
1900	Married Harriet Towle Bradley		'The geography of the region about Devil's Lake and the Dalles of Wisconsin' (with R.D. Salisbury)	
1901	Appointed Instructor in Geology Univ. of Chicago	U.S. Geological Survey (to 1949)		
1903	Ph.D. degree		*Glaciation of the Uinta Mountains* (Ph.D. diss.)	Geography Department established, Univ. of Chicago
1908		Secretary, Chicago Academy of Sciences	'The interpretation of topographic maps' (with R.D.Salisbury)	
1909		Began study of San Juan Mountains	'Glaciation of the Uinta and Wasatch Mountains'	Freud lectures at Clark Univ.
1913	Appointed Professor of Physiography, Harvard Univ.			
1914				First World War
1916				U.S. National Park Service established
1918				End of war
1920	Appointed President, Clark Univ.		*New Geography – Book Two*	
1921	Appointed Director, Graduate School Geography, Clark Univ.	President, National Council of Geography Teachers	Inaugural address: 'The new meaning of geography in American education'; Presidential address National Council of Geography Teachers: 'Geography and world relations'	Graduate School of Geography, Clark Univ. established
1926		Pacific Science Congress, Tokyo. Travel in Japan, Korea, China		

DATES	LIFE AND CAREER	ACTIVITIES, TRAVEL FIELDWORK	PUBLICATIONS	CONTEMPORARY EVENTS
1928		Int. Geog. Cong., Cambridge. Travel in England and Europe	Atwood physical-political wall map series	
1929		President, National Parks Association		
1932		President, Pan American Institute of Geography and History; Carnegie Institute Expedition, to Guatemala; Pan American Inst., Rio de Janeiro; travels in South America 1932-3	'Physiography and quaternary geology of the San Juan Mountains' Colorado, (with K.F. Mather)	
1935		American Scientific Congress, Mexico City	Presidential address, Association of American Geographers: 'The increasing significance of geographic conditions in the growth of nation-states'	
1938		Travel in Southern and Equatorial Africa and Mediterranean	'Working hypothesis for the physiographic history of the Rocky Mountain region' (with W.W. Atwood, Jr.)	
1940			*The Physiographic Provinces of North America*	Second World War
1944	Sc.D. Worcester Polytechnic Inst.	Distinguished Service Award		
1945			*The Rocky Mountains*	End of war
1946	L1.D. Clark Univ. Appointed President Emeritus on his retirement	Pan American Inst. Caracas		
1948		Helen Culver Gold Medal, Geographic Society of Chicago		
1949	Died 24 July at Annisquam, Mass.	Pacific Science Congress, New Zealand; travel in Australia and New Zealand		

Augustin Bernard
1865–1947

KEITH SUTTON

1. EDUCATION, LIFE AND WORK

For the first forty years of the twentieth century
the work of Augustin Bernard dominated the geographical literature on North Africa, a region of paramount
interest to the French school of colonial geography
and of undoubted relevance to wider international
studies of pioneer settlement.   Whereas his North
African orientation may have been prompted by chance,
following from his first university appointment, it
paralleled an established interest in colonial geography initiated as a student in Paris and affirmed
by a pioneering thesis devoted to the Pacific colony
of New Caledonia.

In 1889 Bernard graduated in history and geography from the Sorbonne in Paris.   Four years previously Marcel Dubois had been appointed *maître de
conférences* there, specifically in colonial geography.
As the dedication to him in Bernard's thesis confirms,
his influence was fundamental.   Indeed, Bernard was
one of that group of geographers around Dubois at the
Sorbonne, and around Paul Vidal de la Blache at the
École Normale Supérieure, who reorientated French
geographical studies away from the restrictive limits
of nineteenth-century historical geography.   Methodological acknowledgements were also made by Bernard to
Louis Himly at the Sorbonne, and to non-geographers
Ernest Lavisse and Charles Seignobos.

After graduation Bernard completed only one year
in secondary school teaching at the lycée of Lorient
in Brittany.   A short-lived interest in oceanography
resulted prior to obtaining leave which allowed him to
prepare his theses for the Doctorat ès lettres awarded
in 1895.   His principal thesis, *L'Archipel de la*

Nouvelle-Calédonie, was the second work of colonial
geography inspired by the new discipline of modern
geography.   The first was *Le Sahara* (1893) by Henri
Schirmer, another pupil of Marcel Dubois.   Although
his was the first real geographical study of the
archipelago, Bernard was unable to visit New Caledonia and based his work on various published studies
of its resources and potentialities.   His thesis
included an appraisal of the processes and results of
a new experiment in penal colonization.   The book
remained a major reference work on New Caledonia for
many years, and, naturally enough, Bernard was the
contributor on the islands in H.R. Mill's encyclopaedic *International geography by seventy authors*
(1899).

Bernard's minor thesis, *De Adamo Bremensi Geographo,* with its study of an eleventh-century German
'geographer' whose writings provide an early picture
of Northern Europe, represented a link with traditional nineteenth-century French geography.   It
followed the then required practice in being written
in Latin.

In 1894 Bernard was appointed lecturer on the
geography of Africa at the École Supérieure des
Lettres in Algiers, fifteen years before it became a
university faculty.   His eight years in Algiers
determined his career, as he immediately immersed
himself in the study of this new environment.   Within
a year or two the incessant flow of articles on a wide
spectrum of North African topics began.   His trans-
ference in 1902 to the chair in colonial geography
(established 1892) at the Sorbonne, which was renamed
the chair in the geography and colonization of North

Africa, ensured that the rest of his career was to be associated with the study and teaching of North African affairs.   The Paris chair occupied by Bernard until his retirement in 1935 at the age of seventy was funded by the colonial governments of Algeria and Tunisia.

Along with his teaching at the Sorbonne Bernard offered courses on North Africa intermittently at the École Coloniale, the École des Hautes Études Commerciales and the École des Sciences Politiques, thus extending his influence to non-geographers.   While still in Algiers in 1900 he gave courses on (a) the Populations of the Northern Sahara (open to the general public), (b) the Geography of Muslim Countries, especially Algeria and Tunisia, and (c) General Geography.   In 1921, in Paris, his courses were (a) French Colonization in North Africa, its methods and results, and (b) Questions of African Geography.   By 1930 they consisted of (a) the Geography of North Africa, (b) the Sahara and Saharan Questions, and (c) practical classes for the Certificate in Colonial Geography and for the *diplôme d'études*.   The consistent thread of North African studies was thus maintained throughout his teaching career.   In 1930 his teaching colleagues at Paris included Emmanuel de Martonne, Albert Demangeon and André Cholley.

Although Bernard's teaching career was spent entirely in two centres, Algiers and Paris, he established wider academic contacts, not least through congresses and excursions.   He attended the Sixth International Geographical Congress in London in 1895 when other French visitors included Lucien Gallois, Jean Brunhes, Paul Vidal de la Blache, and Émile-Félix Gautier.   At the 1925 International Geographical Congress in Cairo he was a member of the French government delegation and he contributed to the 1931 Congress in Paris by leading, with Marcel Larnaude, an excursion to Algeria and the Algerian Sahara.   He had earlier been a member of the 1924 French inter-university excursion to Morocco.

By the 1920s Bernard was a recognized expert on North Africa, the author of several textbooks on Morocco and Algeria, and was working on his *Géographie universelle* volumes.   These were the culmination of the later period of his career when he excelled in collective works evaluating the state of knowledge at a point in time.   Three separate studies on *L'Algérie* in 1929, 1930 and 1931, and the seventh edition of his *Le Maroc* in 1931, represented this later phase.   A link with his earlier field research was provided by the periodic publication, with R. de Flotte de Roquevaire, of the *Atlas d'Algérie et de Tunisie* in 15 parts between 1923 and 1936.   This annotated atlas was conceived in 1910 when the Algerian and Tunisian governments granted financial backing but its eventual incompleteness and the slowness of production was a source of regret to Bernard.   Nevertheless the main sheets appeared, several of them based on his earlier research into the rural settlement geography of the indigenous populations of Algeria and Tunisia, on which monographs were published in 1921 and 1924.   Other major research topics investigated by Bernard concerned the history of Saharan exploration, with Commandant Louis Lacroix, an army officer serving with the Algerian Service des Affaires Indigènes, the evolution of nomadism in

Algeria, again with Lacroix, and the political geography and history of the border between Algeria and Morocco.   Significant research monographs were published on all these topics between 1900 and 1911.

This lengthy teaching, research, and writing career devoted to North Africa and, in particular, to the establishment and achievement of French colonial influence there, reached an appropriate climax in the 1930 celebrations of the centenary of French rule in Algeria.   Bernard was a member of the Commission des Publications and of the Comité Métropolitain de Propagande set up by the Minister of the Interior to try and involve the whole of France in the centenary celebrations.   His book, *L'Algérie* (1929), partly funded by the Commissariat Général du Centenaire, won the major prize of 10,000 francs ahead of 37 other works, and was awarded the Lauréat du Grand Concours Littéraire du Centenaire de l'Algérie, both by the City of Paris.   Further, Bernard contributed a lecture on 'Un grand Gouverneur de l'Algérie, Jonnart', who had earlier patronized his *Atlas d'Algérie et de Tunisie*.   This was part of a series of lectures on major Algerian colonial figures held at the École des Hautes Études Sociales.   The 1930 meeting of the Congrès des Sociétés Savantes, held in Algiers, was presided over by Bernard, himself a member of the geography section of the Comité des Travaux Historiques et Scientifiques.   Finally, he attended the Congrès de la Colonisation Rurale, along with Jean Brunhes.

2. SCIENTIFIC IDEAS AND GEOGRAPHICAL THOUGHT

Both Jean Despois and Marcel Larnaude observed that Bernard, together with Paul Vidal de la Blache and Marcel Dubois, was one of the first geographers to orientate French geographical studies away from the restrictive approach of nineteenth-century historical geography.   In his New Caledonia thesis he stated that he regarded geography as a synthesis, rather than an encyclopaedic coverage, of a whole range of phenomena.   He saw geography as a deductive science with two distinct operations: firstly to recognize phenomena, and then to seek explanations.   The mere collection of facts was not the role of the true geographer.   The contemporary interest in the characterization and delimitation of regions was noted but rendered unnecessary by the insular nature of New Caledonia, his area of study.   His later classic article on the natural regions of Algeria was to be his main contribution to this important facet of French geography.   His North African work derived strength from his fieldwork, though as a defence of his thesis on New Caledonia he used the comment of Supan that geography can advance not only by personal observations but also by the synthesis of the observations of others.

Two of Bernard's later research monographs on nomadism and on Algerian rural settlement were to receive methodological praise from Jean Brunhes in his *Human geography*.   Brunhes claims that to 'realize the complexity of nomadism one should first read the book by Augustin Bernard and N. Lacroix'.   He incorporates lengthy sections of the monograph in his book and advocates its policy conclusions, favouring pastoralism over cultivation in the unfavourable

climatic areas of steppe and promoting it through
intensification, water point creation and pasture
improvement.   Of Bernard's research into Algerian
rural settlement, Brunhes concludes 'Throughout the
whole of his study the author remains a geographer,
without ever straying into the realms of technics,
linguistics, ethnography, or folklore'.

Apart from occasional asides, Bernard was a prac-
titioner rather than a theorist in geographical methodology.   He pursued his chosen branch of colonial
geography as writer, teacher and participant in
political and governmental committees on colonial
matters.   The approach of this active subdivision of
French geography was summarized by Albert Demangeon
in the introduction to his *L'Empire britannique*
(1923).   Leaving the description of conquests and the
study of regions to the historian and the regional
geographer respectively, the colonial geographer is
concerned with studying the geographical effect of the
contact of two types of peoples associated by the fact
of colonization.   One group is relatively advanced,
well-provided with capital and material resources, and
has the means to exploit them backed up by the necessary spirit of enterprise and adventure.   By contrast,
the other group, constrained by isolation, turned in
upon itself, has clung to outmoded ways of life, with
resultant limited horizons and poor technology.

While Bernard cannot be accused of neglecting the
study of the indigenous group, he certainly concluded
that the colonial contact was overwhelmingly beneficial for those colonized.   After a detailed survey
of the native people of New Caledonia, he concluded
that they were of little use for the work of colonization.   A parallel conclusion regarding the use of
penal colonization, though of transported Europeans,
left free French emigrants as the only reliable base
for the colony's future development.   A similar,
rather extreme judgement was delivered on the impact
of the native-settler contact in Algeria on the eve
of the centenary of French rule in 1930.   France had
pacified, equipped and organized its colony.   The
settlers had revived formerly cultivated areas, added
new crops, reclaimed land, developed its soil and mineral resources and stimulated its commerce and
industry.   France had even created a new French
Algerian people through the fusion of the various
European elements.   Thus, according to Bernard, the
indigenous people should, without any reservations, feel
able to associate themselves with the celebration of a
conquest which had been totally beneficial for them.
This association could then continue to promote the
further development of the country.   A further display
of this imperialistic viewpoint came in Bernard's
statement that 'The success of all colonizing peoples,
from the Romans to the English, rests on the feeling of
the superiority of the imperial race.' (*L'Algérie*, 1929,
p.522)

Perhaps the circumstances of the impending centenary of French rule in Algeria prompted such extreme
positions.   Certainly, after 1930 Bernard appeared to
return to the less chauvinistic central theme of colonial geography, the study of the contact of two peoples
associated by colonization.   His contribution to W.L.G.
Joerg's *Pioneer settlement* of 1932 recognized problems
associated with the settlement of a large European
population in the midst of a native people which would
always outnumber it.   It was problematic to establish
a system of displacement which would not be unjust for
the native people.   He advocated the approach of
official colonization, though he said little on the
acquisition of land for redistribution.   This policy
should even be carried out in Morocco and Tunisia.
The resulting presence of Europeans in rural areas
would bring the native people closer to France and
wean 'them from their ancient barbarism'.   While a
degree of co-operation from the native people should
be sought, 'as many European families as possible
(should) be dispersed among the natives to act as
their counsellors and their monitors.'

The cruder elements of paternalism appear to have
been dropped from Bernard's approach to colonial geography by the time his two volumes in the *Géographie
universelle* series were published.   In West Africa,
in particular, he recognized that many disastrous
consequences for the indigenous people resulted from
early contacts with European traders and slavers.
While still writing in terms of race, rather than of
culture or civilization, he recognized that forced
assimilation was likely to be unsuccessful and that
'white civilization' might not be appropriate for a
black people, certainly without a more gradual evolution accomplished by the indigenous group itself.
Nevertheless, in rural North Africa he still considered that increased French settlement was needed
for the advancement of the natives.

3. *INFLUENCE AND SPREAD OF IDEAS*

As occupant of a chair at the Sorbonne between 1902
and 1935, Bernard taught and influenced generations
of French geographers.   His wider teaching commit-
ments, with courses on North Africa and colonial geography in several of the Écoles in Paris, considerably broadened both his own and geography's influence
in political, commercial and administrative circles.
The strength of colonial geography in France in the
1930s and 1940s owed much to Bernard.   The process
of European colonization formed a section in several
geographical studies of parts of the French and other
colonial empires.   One can instance Charles
Robequain's books on Madagascar and on South-East
Asia, Hildebert Isnard's studies on Madagascar,
Réunion, and Algeria, Jean Poncet's on Tunisia, and
especially Jean Despois's two tomes specifically
focused on European colonization in Libya and in
Tunisia.   In addition to French colonial geographers,
Walter Fitzgerald, in his *Africa*, made particular
reference to Bernard in the chapter on 'The Barbary
States'.   Bernard retired in 1935 and was succeeded
by Marcel Larnaude.   In 1936 a second chair was
established in the University of Paris, using once
again the title 'Colonial Geography', which had been
discarded in 1902.   This chair was given to Charles
Robequain.   Also, in 1946, three new chairs of
colonial geography were created in France: at Strasbourg, held by Jean Dresch, at Aix-en-Provence, by
Hildebert Isnard, and at Bordeaux by Eugène Revert.
Moreover, the first two of these geographers were and
remain North Africanists.

The influence of Bernard's writings extends over
a long period stretching from his early publications
to his major post-retirement volumes in the *Géographie*

universelle series.   For the first three decades of
the twentieth century, much of the progress made in
developing the geographical knowledge of North Africa
can be attributed to Augustine Bernard and Émile-Félix
Gautier, both of whom devoted the bulk of their
research and writing to the region.   Bernard's 1902
study of the natural regions of Algeria made for the
first time the basic distinction between the coastal
mountain chains and interior chains of the Atlas.
Later, in writing his 1937 volume on North Africa, he
relied more on Gautier's studies of its physical geography but elsewhere in the book he developed his own
ideas on the area's geographical regions -- an immense
pioneering effort given the lack of regional monographs (many of which came later) and the absence of
strongly individualized *pays* already identified by
their inhabitants such as existed in Europe.   Territories were designated by the names of the tribal
groupings occupying them rather than by long-estab-
lished *pays* names.   The resulting arbitrariness of
some of Bernard's regional divisions do not detract
however from the authority of this first all-embracing
geographical study of North Africa to be published
since Élisée Reclus's contribution in his *Géographie
universelle* fifty years previously.

Three distinct levels of influence of Bernard's
writings can be suggested: other geographers, other
specialists in colonialism, and the general public.
Contemporary geographers of wider methodological
orientation, such as Jean Brunhes, made laudatory use
of Bernard's researches into nomadism and rural
settlement patterns.   Other North Africanists quar-
ried his early research monographs and the successive
sections of the *Atlas d'Algérie et de Tunisie*, and
thereafter made contributions.   No discussion of
physical regions in Algeria could ignore Bernard's
suggested divisions.   Bibliographically, contemporary
specialists in North African geography were kept
informed of recent publications through his repeated
contributions to the *Bibliographie géographique
annuelle* (later *internationale*), wherein he succinctly
summarized publications in a wide spectrum of journals.
Geographical studies on all or part of North Africa,
with the strange exception of Émile-Félix Gautier's
Le Sahara in 1928, made frequent reference to the many
articles and books of Bernard.   The French tradition
of providing major geographical studies of the region
made his publications useful to several authors.
Pierre Birot and Jean Dresch stressed several of his
works in their *La Méditerranée et le Moyen-Orient*
(1953), as did Jean Despois in the 1949 and subsequent
editions of his *L'Afrique du Nord* and Walter Fitzgerald in his *Africa*.

From the outset Bernard sought a wider audience
than just geographers.   In particular he published
in journals devoted to colonial administrative and
commercial interests.   Around 1900 he directed the
journal *Questions Diplomatiques et Coloniales*.   Under
his influence a series on the colonial role of French
towns and regions, such as Marseilles, was published:
many of these were written by geographers.   As a member of the Comité de l'Afrique Française, he regularly
contributed to its journal, *L'Afrique Française*, to
the *Comptes Rendus de l'Académie des Sciences Coloniales*, and occasionally to several other political
journals.

The wider reading public valued Bernard's
writings on Algeria and Morocco and his *Le Maroc* went
through seven editions from 1913 to 1931.   The
fourth and fifth editions of M. Wahl's book *L'Algérie*
were edited and updated by Bernard in 1903 and 1908
respectively, and were then replaced by his own book
in 1929, to be followed by two other books on that
country by 1931.   At an even wider level he contributed to four *Guides-Joanne* on Algeria and Tunisia,
1903-9, and to several collective popular works on the
French colonial empire.

This active participation and publishing in both
academic and public life was recognized by a succession of awards and appointments.   Bernard was Sec-
retary-General and technical adviser to the Commission
Interministérielle des Affaires Musulmanes from 1918
until 1936.   The 1930 celebrations for the centenary
of Algeria's status as a French colony brought him
membership of the publications commission and the
Paris-based Comité Métropolitain de Propagande.   His
study of *L'Algérie*, written for the celebrations, was
awarded the Lauréat du Grand Concours Littéraire du
Centenaire de l'Algérie, a literary prize open to all
disciplines.   As a member of the Académie des
Sciences Coloniales and of the Comité de l'Afrique
Française, his career was crowned by membership of the
Académie des Sciences Morales et Politiques in 1938.

Several of these appointments were not just
deserved recognition of earlier achievements, but
involved duties and activities, as was especially
the case with his work for the Commission Interministérielle and the Comité de l'Afrique Française.
Even before the establishment of the French protector-
ate in 1912, he participated in several official
missions to Morocco and was active in campaigning for
its inclusion in the French colonial sphere.   To this
end he prepared a pamphlet on Morocco's economic value
for the use of chambers of commerce and businessmen.
Bernard was particularly closely linked with the
Comité de l'Afrique Française, an essentially political organization pursuing nationalist and imperialist
aspirations.   Through these two organizations he was
able to promulgate further his views on the evolution
of political ideas in North Africa and to obtain an
influential audience for his opinions about the
official colonization of rural areas.

By the 1930s, however, while colonial attitudes
as expressed by people such as Bernard were still in
the mainstream of contemporary political thought,
religious and political nationalism had emerged in
Algeria as elsewhere in Africa.   The advocacy of
assimilation by Ferhat Abbas, especially in the 1930s,
might well have been welcomed by Bernard but the
aspirations of many Algerians were better met by the
religious-cultural movement of Abdulhamid Ben Badis
or by the popularist socialism of Messali Hadj, both
of which contributed to the withdrawal of the French
from their colonial hegenomy of North Africa.   The
more positive interpretation of colonial rule advocated by Bernard and by some other French colonial
geographers was soon to be rendered obsolete.

Bibliography and Sources

1. REFERENCES ON AUGUSTIN BERNARD AND COLONIAL GEOGRAPHY

Larnaude, Marcel, 'Augustin Bernard (1865-1947)', *Ann. Géogr.*, vol 57 (1948), 56-9

Despois, J., 'Augustin Bernard (1865-1947)', *Rev. Afr.*, vol XCII (1948), 217-24

Larnaude, M., 'Émile-Félix Gautier (1846-1940) et Augustin Bernard (1865-1947)', *Bull. Sect. Géogr. Com. Trav. Hist. Sci.*, vol LXXXI (1968-74), 107-18

Church, R.J. Harrison, 'The case for colonial geography', *Trans. Inst. Br. Geogr.*, vol 14 (1948), 17-25

2. SELECTIVE AND THEMATIC BIBLIOGRAPHY OF WORKS BY AUGUSTIN BERNARD

a. Colonial geography

1895 *L'Archipel de la Nouvelle-Calédonie*, Paris, 458p

1899 'New Caledonia', in Mill, H.R. (ed.), *The international geography by seventy authors*, London, 644-6

1900 'La main d'oeuvre aux colonies', *Quest. Dipl. et Colon.*, vol X, 333-50

1919 'La Syrie et les Syriens', *Ann. Géogr.*, vol 28, 33-51

b. North Africa

1898 *L'atlas marocain d'après les documents originaux par P. Schnell* (translated by A. Bernard), Paris, 316p.

1898 'Hautes-plaines et steppes de la Berbérie', *Bull. Soc. Géogr. et Archéol. d'Oran.*, vol 20, 18-31

1899 'La question du Transsaharien', *Quest. Dipl. et Colon.*, vol 8, 10-19

1900 (with N. Lacroix) *Historique de la pénétration Saharienne*, Algiers, 187p.

1901 'L'Afrique du Nord et 1'Empire colonial français', in C. Mourey and L. Brunel, *L'année coloniale - deuxième année 1900*, Paris, 3-20

1902 (with E. Ficheur) 'Les régions naturelles de l'Algérie', *Ann. Géogr.* vol 11, 221-46, 339-65 and 419-37

1902 'L'Algérie. Le pays. La mise en valeur', and 'Le Sahara. Le pays. L'administration', in M. Petit, *Les colonies françaises*, vol 1, Paris, 163-76, 289-302, 471-8, 506-8

1903 M. Wahl, *L'Algérie*, 4th edition updated by A. Bernard, Paris, 454p.

1904 'Une mission au Maroc. Rapport à M. le Gouverneur-général de l'Algérie', *Renseign. Colon. et Doc. Com. Afr. Fr.*, vol 14, 221-43, 258-74

1906 (with N. Lacroix) *L'évolution du nomadisme en Algérie*, Algiers and Paris, 342p.

1908 M. Wahl, *L'Algérie*, 5th edition updated by A. Bernard, Paris, 454p.

1909 'L'irrigation en Algérie-Tunisie', in Institut Colonial International, *Les différents systèmes d'irrigation*, vol 4, 11-48

1909 'Les ressources économiques du Maroc', and 'La propriété immobilière au Maroc', *Congr. de l'Afr. du Nord 1908. C.R. des Trav.*, Paris, 611-20 and 740-50

1910 'L'oeuvre française dans les confins algéro-marocains et ses résultats politiques', *Renseign. Colon. et Doc. Com. Afr. Fr. et Com. Maroc.*, vol 20, 381-91

1911 'Le "dry-farming" et ses applications dans l'Afrique du Nord', *Ann. Géogr.*, vol 20, 411-30

1911 *Les confins algéro-marocains*, Paris, 420p.

1913 'L'Algérie et la Tunisie', in A. Bernard *et al.*, *L'Afrique du Nord*, Paris, 276p, 7-33

1913 *Le Maroc*, Paris, 412p.

1916 *L'effort de l'Afrique du Nord*, Paris, 32p.

1917 (with E. Doutté) 'L'habitation rurale des indigènes de 1'Algérie', *Ann. Géogr.*, vol 26, 219-28

1918 'L'organisation communale des indigènes de l'Algérie', *Rev. Polit. et Parlementaire*, 36p.

1921 *Enquête sur l'habitation rurale des indigènes de l'Algérie*, Algiers, 151p.

1921 *Le régime des pluies au Maroc* (Mémoire de la Soc. des Sciences Naturelles du Maroc), Paris, 95p.

1922 'Un nouveau peuple: l'Algérie et les Algériens', *Rev. Fr.*, vol 2, 580-97

1923 (with R. de Flotte de Roquevaire) *Atlas d'Algérie et de Tunisie*, fascicule 1, Algiers and Paris (15 parts published 1923-36)

1924 *L'habitation rurale des indigènes de la Tunisie*, Tunis, 103p.

1924 (with P. Moussard) 'Arabophones et berbérophones au Maroc', *Ann. Géogr.*, vol 33, 267-82

1926 'La charrue en Afrique', *C.R. Congr. Int. Géogr. Cairo*, vol 4, 283-93

1926 *L'Afrique du Nord pendant la guerre*, Paris, 163p.

1926 (with Dr Paul Vermale) *Au Sahara pendant la guerre européenne. Correspondance et notes.* Paris, 233p.

1927 Ch. Féraud, *Annales tripolitaines*, introduction and notes by A. Bernard, Tunis and Paris, 478p.

1929 *L'Algérie*, Paris, 522p.

1930 *L'Algérie*, 'Histoire des colonies françaises' series, Paris, 548p.

1931 *L'Algérie*, 'Collection coloniale' series, Paris, 224p.

1931 *Le Maroc*, 7th edition, Paris, 481p.

1931 (with M. Larnaude), *Congr. Int. Géogr. Paris 1931. Excursion B4. Algérie, Sahara algérien,* Paris, 48p.

1932 'Rural colonization in North Africa', in W.L.G. Joerg, *Pioneer settlement*, American Geographical Society Special Publication no. 14, New York, 221-35

1937 *L'Afrique septentrionale et occidentale*, part 1, (vol 2 of the *Géographie universelle* series), Paris, 282p.

1939 *L'Afrique septentrionale et occidentale*, part 2, 245p.

c. France

1892 'L'île de Groix', *Ann. Géogr.*, vol 1, 259-78

1923 *Le Bourbonnais et le Berry. Choix de textes, précédé d'une étude*, Paris, 240p.

1938 'Le Prince de Talleyrand à Bourbon-l'Archambault', *Bull. Soc. d'Émulation du Bourbonnais*, 16p.

d. Miscellaneous

1892 'Océanographie: Généralités', *Ann. Géogr.*, vol 1, 199–217
1893 'Océanographie: Océan Pacifique et Océan Indien', *Ann. Géogr.*, vol 2, 151–72
1893 'Les récifs de coraux', *Ann. Géogr.*, vol 2, 281–95
1894 (with M. Dubois) *Géographie générale. Amerique*, Paris, 320p.
1895 *De Adamo Bremensi Geographo*, Paris, 104p.

Keith Sutton is a lecturer in geography at the University of Manchester.

CHRONOLOGICAL TABLE: AUGUSTIN BERNARD

DATES	LIFE AND CAREER	ACTIVITIES, TRAVEL, FIELDWORK	PUBLICATIONS	CONTEMPORARY EVENTS AND PUBLICATIONS
1865	Born at Chaumont-sur-Tharonne (Loir-et-Cher)			
1881				Tunisian Protectorate declared
1885				Marcel Dubois appointed *Maître de conférences* in colonial geography at Paris
1889	Graduated in history and geography			
1890	Lycée teacher at Lorient (Brittany)			
1892				First chair of colonial geography instituted at Paris; holder Marcel Dubois
1894	Lecturer on the geography of Africa at the École Supérieure des Lettres in Algiers			
1895	Doctorat ès lettres	Intern. Geogr. Congr., London	*L'Archipel de la Nouvelle-Calédonie*	
1900		Directed journal *Questions Diplomatiques et Coloniales*		
1902	Appointed to chair of geography and colonization of North Africa at the Sorbonne, Paris		*Les régions naturelles de l'Algérie*	
1906			*L'évolution du nomadisme en Algérie*	Algeciras conference over future of Morocco
1908				Congrès de l'Afrique du Nord in Paris
1909				École Supérieure de Lettres d'Alger became a university faculty
1910		M. Jonnart, Governor-General of Algeria, assured financial assistance for proposed *Atlas d'Algérie et de Tunisie*		
1911			*Les confins algéro-marocains*	Italian invasion of Libya
1912				Moroccan Protectorate established
1913			*Le Maroc*	
1918-36		Secretary-General of the Commission Interministérielle des Affaires Musulmanes		

DATES	LIFE AND CAREER	ACTIVITIES, TRAVEL FIELDWORK	PUBLICATIONS	CONTEMPORARY EVENTS AND PUBLICATIONS
1920		Inter-university excursion in Algeria—prevented by illness from leading it		
1921			*Enquête sur l'habitation rurale des indigènes de l'Algérie*	
1921-7				Rif revolt under Abd el Krim in Spanish Morocco
1923-36			First part of the *Atlas d'Algérie et de Tunisie*	
1924		Inter-university excursion to Morocco - a member		
1925		Intern. Geogr. Congr., Cairo. Member of French government delegation. Contributed paper		
1926				Algerian nationalist party Étoile Nord-Africaine, founded by Messali Hadj
1928				E.-F. Gautier, *Le Sahara*
1929			*L'Algérie*	
1930	Awarded prize for *L'Algérie*		*L'Algérie* - a different study in a colonial history series	Centenary of French Algeria
1931		Intern. Geogr. Congr., Paris. Led excursion with Larnaude to Algeria	*Le Maroc*, 7th edition	
1932			'Rural colonization in North Africa', chapter in W.L.G. Joerg, *Pioneer settlement*	
1934				Pacification of Morocco achieved Nationalist party started in Tunisia by Habib Bourguiba
1935	Retired from Sorbonne			Marcel Larnaude succeeded Bernard in chair at Paris
1936			Last part of unfinished *Atlas d'Algérie et de Tunisie* published	Chair of colonial geography re-established at Paris - Charles Robequain appointed Blum reforms defeated by French Algerians - would have granted limited political rights to Muslims

DATES	LIFE AND CAREER	ACTIVITIES, TRAVEL, FIELDWORK	PUBLICATIONS	CONTEMPORARY EVENTS AND PUBLICATIONS
1937			*L'Afrique septentrionale et occidentale,* part 1	
1938	Entered Académie des Sciences Morales et Politiques			
1939			*L'Afrique septentrionale et occidentale,* part 2	
1940	Retired to Bourbon-l'Archambault (Allier)			German invasion of France
1943				Liberation of Algeria from pro-Vichy régime
1946				Three new chairs of colonial geography created at Strasbourg, Aix-en-Provence and Bordeaux
1947	Died 29 December at Bourbon-l'Archambault			

Antoine d'Abbadie

1810–1897

MIREILLE PASTOUREAU

1. EDUCATION, LIFE AND WORK

Antoine d'Abbadie, also known as Thompson d'Abbadie, was born in Dublin to an Irish mother and a French father who had left his native Basse-Pyrénées at the beginning of the French Revolution and did not return home with his family until 1813. This Basque-Irish origin may in part explain d'Abbadie's determination and perseverance of character, his devotion to religion, his love for the Basque countryside and language (on which he wrote several studies) and an underlying lack of sympathy with the British which became apparent during his expedition to Ethiopia.

Antoine d'Abbadie's wide interests extended far beyond those normally found among geographers: as he explained in *Geographie de l'Éthiope* (1890). 'The observations of any visitor, whether as a geographer or as a naturalist, in an unknown or little known country, attracted me far less than the study of languages, laws, religions, political and legal constitutions and literature.' While following his legal studies d'Abbadie also found time to become proficient in mineralogy, geology and botany, and even the construction of scientific instruments.

On leaving college he read Chateaubriand's *Les Natchez* in which, like the famous French naturalist Count Buffon (1707–88), he found the fascination of natural history and of astronomy. So inspired he decided to explore the interior of Africa. His first idea was to travel through Tunisia and Morocco but when he read the works of James Bruce (1730–94) some fifty years after they were written he longed to visit Ethiopia, where the mystery of the source of the Nile might be solved.

Following a visit of a year to Brazil, arranged by François Arago on behalf of the Academy of Sciences to study the effect of the sun on the magnetic needle, Antoine d'Abbadie went to Ethiopia with his brother Arnaud. The two brothers, aged 27 and 22, stayed for 11 years, of which the first few were given largely to accumulating equipment, including compasses, suited to the highly fragmented topography of Ethiopia. The real task of exploration began in 1840 and with it came many problems, including an accidental gunshot wound which left Antoine partially blind, difficulties with the incessantly fighting population and in time the hostility of English explorers. The brothers found it best to work independently, and Arnaud became in effect judge, diplomat and general in the Ethiopian armies. Antoine went to the south where, from Gondar and Lake Tana, he followed the rivers Abbai(the Blue Nile), Gibé and Omo, finally reaching the southern limit of his long and difficult journeys at a point south of the river Gogeb, a tributary of the Omo. The expedition came to an end in 1848 after the third d'Abbadie brother, Charles, had been sent to find Antoine and Arnaud, whose movements had often been difficult to trace. Antoine never explored again, and apart from participation in a few brief scientific missions he lived permanently in his observatory at Chateau d'Abbadie near Hendaye, which he embellished in Ethiopian style. There he spent his time studying astronomy and publishing the results of his exploration.

His scientific missions included visits to Norway in 1857 and to Spain in 1860 to observe solar eclipses. In 1882 he directed the French expedition

sent from Port-au-Prince to observe the transit of
Venus in Haiti and Santo Domingo and in 1884 he travelled at his own expense via Athens, Jerusalem and
Istanbul to Egypt and the Red Sea, where he studied
terrestrial magnetism.

2. *SCIENTIFIC IDEAS AND GEOGRAPHICAL THOUGHT*
Primarily a man of action, Antoine d'Abbadie was not
an original geographical thinker. His one scientific
hypothesis, on the source of the Nile, turned out to
be wrong. But through his long period of exploration, the serious character of his work and the vast
knowledge he acquired, he became the most renowned
explorer of Ethiopia. Before his time Enarea (or
Inarya) was unknown to Europeans. Never before had
there been a reconnaissance survey of the Galla upland
although the area to the east of Lake Tana had been
traversed by Ruppell in 1831-3 and the area to the
south of the lake visited by Combes and Tamisier in
1835-6. Antoine d'Abbadie reached the northern frontier of Kaffa, taking with him the geodetic equipment
needed to make a map. Among his own instruments,
which proved to be accurate as well as portable, was a
new theodolite, which he called the *aba*,. for measuring
heights. In effect he had become a geodetic explorer.

His work on linguistics and ethnology is of
equal significance. A careful investigator in all
his work with the people of Ethiopia, he learnt some
thirty local languages and compiled the first dictionary of the Amharic language, comprising some 40,000
words. He also planned a vast study of the population, and listed all the personal names of the
inhabitants in five villages, as well as compiling a
gazetteer of 6,000 inhabited places. His ardent
Catholicism impelled him to petition Pope Gregory XVI
to evangelize Ethiopia and the first of three missions
came in 1839. He collected more than 250 Ethiopian
manuscripts and bequeathed them to the Academy of
Sciences, which in time sent them to the National
Library. He was equally interested in coins and
published a numismatic catalogue with his catalogue of
manuscripts. Also found after his death was a vast
quantity of notes on history, philology and ethnography, which became a main source for the work of the
Italian expert on Ethiopia, Carlo Conti Rossini (1872-
1948).

3. *INFLUENCE AND SPREAD OF IDEAS*
Antoine d'Abbadie was regarded as one of France's
greatest explorers. The Société de Géographie gave
him a silver medal in 1839 and a gold medal in 1850.
Recognition came also from the Académie des Sciences,
who appointed him correspondent for the geographical
and navigational section in 1852, a full member in
1867 and a member of the Bureau des Longitudes in 1878.

Despite all these enviable honours, his misappre-
hension as to the source of the Nile was a blight on
his career. In January 1844 he announced triumphantly that the river Gogeb was the main contributor to the White Nile of which, in April of the same
year, he claimed to have found the source. However,
in the *Bulletin de la Société de Géographie* from 1845
a fierce controversy arose provoked by the English

explorer Charles Tilstone Beke (1800-74). D'Abbadie
put up a vigorous defence but eventually realized he
was mistaken and became reluctant to publish the
results of his explorations. Deprived of the kudos
of making a great discovery he was content to become
an adviser on travel and exploration. Some of his
published work was greatly appreciated, notably his
map of Ethiopia in ten sheets, 1862-9, his *Géodésie
d'une partie de la Haute-Éthiopie*, a major work published from 1860-3, the *Observations relatives à la
physique du Globe* of 1873, and his *Dictionnaire de la
langue Amarinna* of 1881.

Bibliography and Sources

1. *REFERENCES ON ANTOINE D'ABBADIE*
Anon., *Notice sur les travaux scientifiques de M.
    Antoine d'Abbadie*, Paris (1854), 18p. and also
    Paris (1867), 25p.
Nansouty, M. de, 'Antoine d'Abbadie', *Bull. de la Soc.
    Ramond* (1897), 45-51.
Thirion, J., 'Antoine d'Abbadie', *Rev. des Questions
    Sci.*, Louvain (April 1897), 589-607
Hatt, M., 'Notice sur la vie et les travaux de M.
    d'Abbadie', *C.R. Acad. des Sci.*, Paris (17 Jan.
    1898), 173-81.
Deherain, H., 'Antoine d'Abbadie explorateur de
    l'Ethiopie', *Études sur l'Afrique*, Paris (1904),
    107-19.
Darboux, G., *Notice historique sur Antoine d'Abbadie,
    lu dans la séance publique annuelle du 2
    décembre*, Inst. de France, Acad. Sci., Paris
    (1907), 42p.
Froidevaux, H., 'Antoine d'Abbadie', *Dict. Biog. Fr.*,
    Paris (1932), vol 1, 35-42
The controversy on the sources of the Nile is treated
in Beke, C.T., *A letter to M. Daussy*, London (1850), 8p.
and *An enquiry into M. Antoine d'Abbadie's journey to
Kaffa in the years 1843 and 1844 to discover the
source of the Nile*, London (1850), 56p. Also of
interest is Ayrton, F., 'Observations upon M.
d'Abbadie's account of his discovery of the sources
of the White Nile...', *J.R. Geogr. Soc.*, vol 18
(1848), 48-74.
A contribution of great significance is the book by
Arnaud d'Abbadie, *Douze ans dans la Haute Éthiopie*,
Paris (1868), 637p. Only vol 1 appeared: comment
on vol 2 appears in Tubiana, J., 'Deux fragments
inédits du tome second de Douze ans dans la Haute-
Éthiopie', *Rocznik orientalistyczny*, Warsaw, vol 25/2
(1961), 27-85.

2. *SELECTIVE BIBLIOGRAPHY OF WORKS BY ANTOINE
    D'ABBADIE*
Only the main sources are listed here. The *Bull.*

Soc. *Géogr.* from 1839 to 1848 published a number of letters sent by Antoine d'Abbadie to Jomard and Daussy during the expedition to Ethiopia and there are a number of articles in the *J. Asiat.* and in *Nouv. Ann. des Voyages.*

1839 'Voyage en Abyssinie', *Bull. Soc. Géogr.*, sér 2, vol 11, 200-17

1839 'Sur les sources les plus intéressantes à consulter par les voyageurs', *Bull. Soc. Géogr.*, sér 2, vol 11, 338-40

1849 'Note sur le haut Fleuve Blanc', *Bull. Soc. Géogr.*, sér 3, vol 12, 144-61

1858 *Sur la tonnerre en Ethiopie*, Mémoires présentés par divers savants à l'Académie des Sciences, Paris, vol 16, 158p.

1859 *Catalogue raisonné des manuscrits éthiopiens appartenant à Antoine d'Abbadie*, Paris, 236p.

1859 *Résumé géodésique des positions déterminées en Ethiopie*, Leipzig, 28p.

1860-3 *Géodésie d'une partie de la Haute-Ethiopie... revue et rédigée par Rodolphe Radau*, Paris, 456p.

1866 *L'Arabie, ses habitants, leur état social et religieux, à propos de la relation du voyage de M. Palgrave*, Paris, 75p.

1867 'Instructions pour les voyages d'exploration', *Bull. Soc. Géogr.*, sér 5, vol 13, 257-93

1868 'Observations sur les monnaies éthiopiennes', *Rev. numismatique*, nouv sér, vol 13, 45-60

1868 'L'Abyssinie et le roi Théodore', *Le Correspondant*, Paris, nouv sér, vol 37, 281-321

1873 'Sur un voyage en Éthiopie', *Congr. Sci. de France*, sess 39, vol 1, 121-32

1873 *Observations relatives à la physique du globe, faites au Brésil et en Éthiopie*, ed. M. Radau, Paris, 198p.

1877 'Les causes actuelles de l'esclavage en Éthiopie', *Rev. des questions sci.*, vol 2, 5-30.

1880 'Préparation des voyageurs aux observations astronomiques et géodésiques', *Bull. Soc. Géogr.*, sér 6, vol 20, 75-8

1880 'Sur les Oromo, grande nation africaine désignée souvent sous le nom de Galla', *Ann. de la Soc. Sci. de Bruxelles*, an 4, 1879-80, 167-92

1881 *Dictionnaire de la langue Amarinna*, Actes de la Société philologique, vol 10, Paris, 1336p.

1884 'Exploration de l'Afrique équatoriale: "credo" d'un vieux voyageur', *Atti del Terzo Congr. Géogr. Int.*, Venice (1881), vol 2, 15p.

1890 *Géographie de l'Éthiopie, Ce que j'ai entendu, faisant suite à ce que j'ai vu*, vol 1, Paris, 457p. The second volume never appeared.

3. UNPUBLISHED AND OTHER SOURCES ON ANTOINE D'ABBADIE

A number of manuscript sources have not been published. Antoine d'Abbadie died childless and left most of his papers to the library of the Institut de France: of these a catalogue has been made by Gaston Daroux. The eighteen boxes of manuscript notes made during his expedition are in the manuscript department of the Bibliothèque Nationale and have been partially published in Tubiana, J., 'Fragments du journal de voyage d'Antoine d'Abbadie', 30p. in no. 5 (1959) of *Mer Rouge, Afrique orientale, Cahier l'Afrique et l'Asie.*

Several letters are preserved in the archives of the Société de Géographie, and in the Department of maps and plans of the Bibliothèque Nationale there are a number of manuscript maps as well as the corrected proofs of the map of Ethiopia.

Mireille Pastoureau is a curator in the Département des Cartes et Plans at the Bibliothèque Nationale, Paris. Translated by T.W. Freeman.

CHRONOLOGICAL TABLE: ANTOINE D'ABBADIE

DATES	LIFE AND CAREER	ACTIVITIES, TRAVEL, FIELDWORK	PUBLICATIONS	CONTEMPORARY EVENTS
1810	Born in Dublin, 3 January			
1813	Returned with d'Abbadie family to France			
1827	Began law studies at Toulouse: these were continued in Paris from 1828 when his family went there to live permanently			
1836		Visit to Brazil, Académie des Sciences		
1837		Travelled to Ethiopia with his brother Arnaud		
1838				Egyptian expedition to Ethiopia
1839	Silver medal awarded by Société de Géographie, Paris	Returned to France to purchase scientific instruments		First of three Catholic missions to Ethiopia
1840		Antoine and Arnaud d'Abbadie decided to travel independently; in November Antoine went to Lake Tana		Ethiopia divided into four states
1843		Antoine was the first European to enter Inarya		
1844		Study of the river Omo and its tributaries by Antoine	In April, Antoine announced that he had discovered the source of the White Nile	
1848		End of Ethiopian expedition		
1849			'Note sur le haut Fleuve Blance', *Bull. Soc. Géogr.*	
1850	Gold medal and cross presented to the two brothers by the Société de Géographie			
1851		Visit to Norway		
1852	Corresponding member, Académie des Sciences (geography and navigation section)			
1859	Married Virginia Vincent de Saint-Bomet			

DATES	LIFE AND CAREER	ACTIVITIES, TRAVEL, FIELDWORK	PUBLICATIONS	CONTEMPORARY EVENTS
1860		Visit to Spain	*Géodésie... de l'Éthiopie* also *Carte d'Éthiopie* (to 1863)	Speke and Grant discovered the sources of the Nile
1867	Full member of the Académie des Sciences			
1873			*Observations relatives à la physique du globe*	
1878	Member of the Bureau des Longitudes			
1881			*Dictionnaire Amarinna*	
1882		Visit to Haiti		
1884		Visit to Red Sea		
1890			*Géographie de l'Éthiopie* (never completed)	
1897	Died 20 March, no family			

Alexandre Dimitrescu-Aldem

1880–1917

PETRE COTET

Alexandre Dimitrescu-Aldem was one of the group of pupils of Simion Mehedinţi who began the modern teaching of geography in Romania, among whom were G. Vâlsan, C. Bratescu and N. Orghidon. Although he died at 37, Aldem (as he was generally known) was considered to be one of the most remarkable Romanian geographers of his time.

EDUCATION, LIFE AND WORK
Born in the village of Buhuşi, Moldavia, he was a favoured pupil of Mehedinţi, who inspired him with enthusiasm for geography and chose him as an assistant while he was still a student. He was given a bursary by the Ministry of Education to continue his studies in Germany from 1909-11 at the universities of Göttingen and Berlin. He wrote his doctorate thesis (*Die untere Donau zwischen Turnui Severin und Brăila*) under Albrecht Penck and attended lecture courses and field excursions given by W.M. Davis. Aldem was one of the first geographers in Romania to specialize in geomorphology, but he was trained in the tradition of S. Mehedinţi to become a complete geographer, studying all aspects from cartography to economic geography. He was also a serious student of theoretical geography and methodology.

In an essay on geographical methodology published in 1915, intended for teachers in secondary education, he gave a general view of his conception of modern geography. It included chapters on tectonic movements, relief forms, climate, structure and notably on the Davisian cycle of erosion, a concept which he introduced to Romanian colleagues.

Much earlier, in 1903, only two years after he had taken his first degree, his paper on 'Man and environment in the Carpathians' dealt with the relationship between man and his environment and was followed by other papers in human geography. These included studies of the village of Ocina in 1905, of the density of population in Moldavia in 1909 and of Romania's position in southeast Europe. He also wrote on cartography and geomorphology. For the Institute of Military Geography he provided ethnographic maps, which were never published, showing the proportion and distribution of Romanians in Transylvania and Bukovina; this became an issue of significance towards the end of the First World War.

Although Aldem never lost his interest in human geography, in methodology (especially in relation to teaching), geographical theory (above all in geomorphology) and cartography, his German experience gave him an absorbing interest in geomorphology. His doctorate thesis was the first contribution from any Romanian author on the geomorphological analysis of the lower Danube. Based on a detailed study of the river terraces, his conclusion was that the lower Danube was of Quaternary formation, becoming increasingly recent in date downstream.

Having published his thesis in 1911, Aldem continued his researches on the Danube, on which G. Vâlsan also wrote a thesis. Aldem held the view that the lower Danube plain could be divided into three sections: in the lower reaches, to the east of Arges, there was a depression; between Arges and Jiu there existed some elevation in Quaternary times; while to the west of Jiu there was a state of equi-

librium.   These views were disputed by Vâlsan, but
recent researches, such as those of Brătescu and
Coteţ, have at least partially confirmed Aldem's
views.

Aldem died before the end of the First World War
and has been almost forgotten, eclipsed by the renown
of his fellow-geographers G. Vâlsan and C. Brătescu.
Nevertheless, he was a geographer with both a wide
range of learning and a special interest in modern
geomorphology based on a critical analysis of the
relation between climate and tectonics as genetic
factors in landform evolution.   He inclined to causal
and genetic views in his geographical work and
initiated the writing of monographs on villages.
Though his life was short, he became one of the distinguished Romanian geographers of the early twentieth
century.

1915 'W.M. Davis in literatura geografica contim-
porana', *Bul. Soc. (regale) Rom. de Geogr.*,
vol 35, 175-9

*Petre Coteţ is Professor of Geography at the Univer-
sity of Bucharest.   Translated from French by T.W.
Freeman.*

Bibliography and Sources

1. REFERENCE ON ALEXANDRE DIMITRESCU-ALDEM

Coteţ, P., 'Alexandre Dimitrescu-Aldem', *Terra* 1,
Bucharest (1969), 26-9 (in Romanian with English and
Russian abstracts)

2. PRINCIPAL WORKS OF ALEXANDRE DIMITRESCU-ALDEM

a. Theoretical geography and methodology
1910 *Studiul geografiei în învătămiit* (Geographical
study and teaching), Bucharest
1915 *Cîteva puncte cardinale în ale geografiei* (Some
cardinal points in geography), Bucharest

b. Human and economic geography
1903 *Naekra si omul in Carpati* (Man and environment in
the Carpathians), Bucharest
1905 *Satul şi locuitoru Ocinei* (Ocina village and its
people), Bucharest
1909 *Densitatea populatiei die Moldora* (Population
density of Moldavia), ed Albert Boer, Bucharest

c. Geomorphology
1911 Doctorate thesis (in German) *Die untere Donau
zwischen Turnui Severin und Braila*, Geomorpo-
logische Betrachtungen Akademischen Buchhandlung
von Conrad Skopnik, Berlin
1911 'Uber die Bildung der Alluvialterrassen', *Geogr.
Anzeiger* (Gotha) vol 12, 101-3
1914 'Asupra teraselor aluvionae' ('On alluvial
terraces'), *Bul. Soc. (regale) Rom. de Geogr.*,
vol 34

CHRONOLOGICAL TABLE: ALEXANDRE DIMITRESCU-ALDEM

DATES	LIFE AND CAREER	ACTIVITIES, TRAVEL FIELDWORK	PUBLICATIONS	CONTEMPORARY EVENTS AND PUBLICATIONS
1880	Born at Buhaşi, Moldavia			
1901				S. Mehedinţi becomes Professor of Geography at Bucharest
1903			'Man and environment in the Carpathians'	
1904	Graduated, specializing in geography. Worked as assistant to Professor Mehedinţi			
1905	Taught geography at the Piatra Neamţ grammar school		'Ocina village'	
1909	Went to Göttingen as a student of Albrecht Penck (to 1911)	To Germany	'Population density of Moldavia'	W.M. Davis in Germany
1910			'Geographical study and teaching'	W.M. Davis, *Die erklärende Beschreibung der Landformen*
1911		Continued travel in Germany; met W.M. Davis	Doctorate thesis on the Danube	
1912	Worked at the Meteorological Institute and with Professor Mehedinţi	Assisted at the Institute of Military Geography		
1913			Essay on the work of W.M. Davis	
1914	Taught at the St Sava grammar school, Bucharest			First World War
1915		Controversy with G. Vâlsan about the Romanian Plain	Methodological paper on 'Some cardinal points in geography'	
1917	Died at Bucharest			

Archibald Geikie
1835-1924

W. E. MARSDEN

Archibald Geikie's pre-eminence as a geologist, and perhaps the equivocal position in which physical geography has long found itself, seem to have concealed the importance of his contribution, both as a thinker and a teacher, to the scientific development of geography.

1. EDUCATION, LIFE AND WORK

Archibald Geikie was the son of an Edinburgh businessman who inclined towards musical pursuits. For many years music critic of *The Scotsman*, he provided a cultured home background for his children. Archibald was the eldest of a family of eight and one of his younger brothers, James *(q.v.)*, achieved almost equal fame as a geologist and physical geographer. After three years in a preparatory school, Archibald entered Edinburgh High School in 1845. Here a lifetime's devotion to the classics took root. The most formative among his early years was 1845, in which he made his first visit to the moors and mountains of Scotland, and the discovery of fossils in a limestone quarry near Edinburgh 'stamped his fate' as a geologist. The variety of geology and scenery in the Edinburgh area was a source of continuing inspiration.

In 1850 Geikie began work in a lawyer's office in Edinburgh, but found it 'unspeakably dull'. A tolerant father allowed him to leave and enter Edinburgh University in 1852. Here he studied natural science and classics until straitened family circumstances forced him to leave early. The university experience was crucial, however, and Geikie was soon marked out for distinction. After a visit to Arran in 1851 his

two geological articles in a local newspaper led to an introduction to Hugh Miller, author of *Old red sandstone*, whose friendly stimulus meant much to the youthful Geikie.

Edinburgh in the Victorian era was one of the great intellectual capitals of Europe. Though the confrontations between the disciples of Hutton and Werner, nowhere more intense than in Edinburgh, had become muted by the middle of the century, they remained fresh in the memories of Geikie's mentors. These included James Pillans, Professor of Humanity, much interested in geography and education; and famous scientists such as John Fleming (naturalist), James David Forbes (natural philosopher), and George Wilson (chemist). Through Wilson, Geikie was introduced in turn to the publisher, Alexander Macmillan, and to A.C. Ramsay, then local director of the Geological Survey.

Like most aspiring geologists of his time, Geikie was largely self-taught. He wandered through the countryside with his copy of MacLaren's *Geology of Fife and the Lothians* and his notebook, and at home studied whatever geological papers he could find. His study served him well. After an excursion with Geikie over Arthur's Seat in what proved to be an informal examination, Ramsay agreed to recommend him for appointment to the recently established Geological Survey of Scotland.

Between 1855 and 1869 Geikie mapped large tracts of the Lothians, Fife, Berwickshire, Ayrshire and Renfrewshire. In holiday periods he visited the Western Isles, on which he published his earliest geological papers. His enthusiasm for fieldwork appeared in his first book, *The story of a boulder: or gleanings from the notebook of a field geologist* (1858). A visit with Murchison, the influential Director General of the Geological Survey, to investigate the structures of the north-west Highlands (1860) reinforced his uniformitarian views on the evolution of scenery, and led to a joint paper which became for a time the definitive statement on Highland structures though the interpretations it contained were later proved to be unsound and emerged as a focus of scientific controversy.

During the 1860s Geikie's immense capacity for work bore fruit in a series of important articles, including 'On the phenomena of the glacial drift of Scotland' and 'On modern denudation', both printed in the *Transactions of the Geological Society of Glasgow*. His studies of the landscapes of his homeland culminated in his first great work, *The scenery of Scotland viewed in connection with its physical geology* (1865), after which his interest switched to the impact of vulcanicity on the British landscape. The glacial field was left, perhaps not consciously, to his brother James, who was soon to produce his classic, *The great Ice Age and its relation to the antiquity of man* (1874).

In preparing their texts, both Geikie brothers attached great importance to comparative data from the study of landscapes abroad. Archibald frequently testified to the crucial insights and scientific and emotional inspiration gained from tours of the Auvergne (1861), Arctic Norway (1865), the Eifel and Switzerland (1868), Austria (1869), southern Italy (1870) and, later, western U.S.A. (1879). Here he examined in close and living detail volcanic and glaciated scenery, and spectacular instances of the impact of river erosion on the landscape. Most of these visits resulted in scientific papers.

In 1867 a separate Scottish branch of the Geological Survey was established and Archibald Geikie, with the support of Murchison, was made its first director. During 1868-9 Geikie completed surveys of western Ayrshire and Renfrewshire, the last systematic field mapping he was to undertake. His increasing stature was attested in his election as President of the Geological Section of the British Association at its Dundee and Edinburgh meetings, in 1867 and 1871 respectively.

In 1871 Geikie married Alice Pignatel, of mixed French-English parentage, and was appointed first Murchison Professor of Geology in the University of Edinburgh. Murchison had made it a condition of endowment that Geikie should occupy the first chair. Reflecting on the former but faded glories of the Scottish school of geology, Geikie resolved to restore its image, and gave his inaugural lecture on that topic. He proceeded to build up departmental resources from scratch, introduced new teaching methods, and handed on at least a partially revived tradition to his brother James, who succeeded him as professor in 1882.

In his years at the university, Archibald Geikie established himself as one of the major textbook writers of the century. He produced texts of physical geography and geology for almost all academic levels, capped by the awe-inspiring *Textbook of geology* (1882). Most of his textbooks ran to several editions. Between 1873 and 1922 his science primer *Physical geography* sold over half a million copies. He also published many articles, some of which were included in the collection *Geological sketches at home and abroad* (1882). His talents as a biographer, first seen with his completion of Wilson's unfinished *Memoir of Edward Forbes* (1861), came to fruition in the *Life of Sir Roderick Murchison* (1875).

On Ramsay's retirement in 1882 Geikie was appointed Director General of the Geological Survey and moved to London. During his first year, he made a point of visiting every officer of the staff in England and Ireland 'in the field'. The most sensitive issue he had to deal with was the 'Highland controversy', for the work of Lapworth and others had challenged the Murchison-Geikie interpretation of the structures. He dispatched two of his most skilled field geologists, Peach and Horne, to the area for further investigation. Their findings confirmed the contrary analysis and Geikie retracted his views publicly in *Nature*. Later initiatives included the mapping of the mining districts on a six inch scale, and expansion of the publications policy of the Survey. He supported Murchison's contention that the Geological Survey was a national asset which should remain in being after the first mapping of Britain was complete. Geikie's last labour for the Survey was the completion of two important memoirs, on central and western Fife and Kinross (1900) and eastern Fife (1902), of which the latter appeared after his retirement in 1901.

In the period between 1885 and 1901 Geikie wrote more textbooks, one of which was *The teaching of geography* (1887). Earlier texts were brought up to date with meticulous care, to keep pace with the growth of international scientific knowledge. Other books included a *Memoir of Sir Andrew Crombie Ramsay* (1895), *The founders of geology* (1897), and *The ancient volcanoes of Great Britain* (1897), a two-volume synthesis of half a lifetime's work on the subject. He also edited volume three of Hutton's *Theory of the earth* (1899), never previously published, and continued to pour out articles.

Distinctions and awards were heaped upon Geikie during this time, of which perhaps the most important were a knighthood (1891) for services to the Geological Survey, the presidency of the British Association (1891), and the Royal Medal of the Royal Society (1895). Foreign travels often included attendance at international conferences. 1897 was his *Wanderjahr* in which visits were made to the United States to lecture at Johns Hopkins University, Baltimore; and to Russia for the International Geological Congress at St Petersburg. This was followed by an extensive tour of southern Russia, and a journey home through the Mediterranean, Italy and France. Further public services included membership of the Royal Commission on the water supply of London, a Departmental committee to enquire into the condition of the Ordnance Survey (both in 1892), and a governorship of Harrow, his son's school.

After his retirement Geikie, taking on a new lease of life, became Secretary of the Royal Society (1903) and later its President (1908). Further honours were conferred, crowned by the K.C.B. (1907) and Order of Merit (1913). In 1906 he was made

President of the Geological Society of London and reappointed in 1907 as a person peculiarly worthy of the honour in the centenary year of the society. In 1910 his extraordinary range of talent was marked by his election as President of the Classical Association. His final piece of public service was to act as Chairman of the Royal Commission of Public Inquiry into Dublin University, set up in 1919.

In this last major period of his life, Geikie's main writings were on literary and classical subjects. A collection of previous articles relating to the impact of scenery on history, literature and the imagination were published as *Landscape in history and other essays* (1905). *The love of nature among the Romans* (1912) was in part inspired by his presidency of the Classical Association, and was associated with pilgrimages to Italy to visit the environs of his beloved Roman poets. 'English science and its literary caricaturists in the 17th and 18th centuries' (1914) and 'The birds of Shakespeare' (1916), emerged from addresses given to the Natural History Society of Haslemere, where he spent the last years of his life. *Annals of the Royal Society Club* (1917), in which he had been a leading spirit for nearly a quarter of a century, was another labour of love. His *Memoir of John Michell* (1918), the eighteenth-century Cambridge geologist, resulted from some intriguing discoveries in Royal Society records. Geikie's last work was his autobiography, *A long life's work*, published a few months before his death on 10 November, 1924, in his 89th year.

The later years of Geikie's triumphantly successful life were marred by bereavement. The tragic deaths of his only son, who had seemed destined for a distinguished career in the civil service, in a railway accident in 1910, and of his second daughter in 1915, were grievous blows. His brother James also died in 1915, and his long-invalided wife in 1916. His final years were spent in the care of his eldest daughter.

As a person Geikie was 'always sane, sound, self-possessed and self-centered' (Clarke, J.M. (1925), *Bull. Geol. Soc. Am.*, 36,128-9). He had patience and a prodigious zeal for work, especially for writing. His considerable administrative abilities were demonstrated in his years as Director General of the Geological Survey and as Secretary of the Royal Society. Strahan, a later director of the Geological Survey, regarded the 'keynote of his success as industry directed by sagacity... A clear if somewhat cold judgement controlled his actions, but in his biographical work the coldness was masked by a studied kindliness of expression. Though he made many friends at home and abroad, his sympathies with his fellow-men were somewhat over-shadowed by his love of Nature and passion for work. He did not seek collaboration, nor could he brook criticism.' (Obituary notice in *Q.J. Geol. Soc.*) G.L. Davies (*The earth in decay* 1968, 345) notes the paradox of Geikie's dedication of the radical *The scenery of Scotland* to the arch-conservative Murchison.

While Geikie was often unstinting in his praise for predecessors and subordinates, and for many overseas colleagues, he expressed little in the way of generous public appreciation of the efforts of his peers, with certain exceptions, such as his tribute in *The scenery of Scotland* to the work of Jukes on the valleys of southern Ireland. It seems strange, however, that in his autobiography he makes little mention even of the work of his brother, his colleague in field mapping in the 1860s, and companion on his Rhine journey in 1868, whose reputation at home and abroad was almost as great as his own.

There seems to be no doubt that Geikie was seen as a heavy, pompous and unyielding establishment figure, regarded with respect rather than affection by his fellow-scientists. As one obituarist notes, his range of skills made him a 'deadly critic' (*Proc. Geol. Ass.* 36 191-2, 1925), and his influence was much sought by those seeking high rank in the scientific world. Strahan, however, records Geikie's keen sense of humour and ability as a raconteur, drawing on a fund of stories from his native heath, some of which are collected together in *Scottish reminiscences* (1904).

2. *SCIENTIFIC THOUGHT AND CONTRIBUTIONS TO GEOGRAPHY*

Archibald Geikie was one of the great 'earth scientists' of the nineteenth century, producing classic papers and texts such as those on the old red sandstone of Scotland (1860) and of western Europe (1879); on the glacial drift (1863) and more general landscape evolution in *The scenery of Scotland* (1865) and 'On modern denudation' (1868); and on the history of volcanic activity in the British Isles, a study consummated in his two-volume masterpiece, *The ancient volcanoes of Great Britain* (1897).

Peach and Horne, however, were of the opinion that Geikie's major contributions to geological science came less from his detailed original researches than from his skilful diffusion of the scientific knowledge explosion of his time ('The scientific career of Sir Archibald Geikie', *Proc. R. Soc. Edinburgh*, 45 (1925)). Cadell was savagely critical of some of the claims made by Geikie with regard to his own geological mapping, and suggested that his 'claim to a niche in the Temple of Fame will, it is to be feared, not rest on his faculty of original observation in the field so much as on his capacity for lucid and attractive exposition' (Review of Geikie's autobiography, *Scott. Geogr. Mag.*, 40 (1924)).

Nevertheless, Geikie made a major contribution to the progress of geomorphological thought. At the time he began his professional career, Lyell's 'uniformitarian' interpretation of the earth's physical evolution was in the ascendant, though catastrophic theories still retained some residual support among the older generation of geologists, such as Murchison. Geikie's observations in the Scottish Highlands (1860), the Auvergne (1861) and later in the Rockies (1879), convinced him of the inexorable power of rain, rivers, the sea and ice in moulding the landscape. After contemplating the stupendous gorges of Le Puy district, he records 'a profound respect for the terrestrial agencies of waste' ('Among the volcanoes of central France' in *Geological sketches at home and abroad*, 125-6). He similarly recounts his excitement at his first contact with 'moving ice' in Arctic Norway, caught 'in the very act of doing the work of which I had hitherto only seen the ancient results' (*A long life's work*, reminiscing on his Norway visit,

1865, 108).

As a follower of Lyell, Geikie for some time accepted his 'marine erosion' theory, and his conversion to fluvialism, combined with his powers of exposition, were of critical importance. Chorley, Dunn and Beckinsale (*The history of the study of landforms*, vol 1, 1964) point out that *The scenery of Scotland* was only nominally concerned with that country, and was written 'to justify and establish the fluvial idea as the correct means of interpreting topography'. They add: 'When Ramsay and Jukes finally had to retire through ill-health, it was Geikie who took their place as the principal advocate of fluvialism' (p.407). They similarly regard highly the 1868 paper 'On modern denudation': 'the most effective quantitative demolition of the marine erosion theory ever published' (p.328). G.L. Davies also indicates that this paper, coupled with a contemporary one by Croll, sounded the 'death-knell' of the marine erosion theory (*op. cit.*, 350-1). Chorley, Dunn and Beckinsale in addition quote W.M. Davis's opinion that the 1868 paper had such an impact in Britain (*op.cit.*, vol 2, 1973, 336-7). In the glacial field, Geikie's work could also be seen as paving the way for the acceptance in Britain of Agassiz' view of the efficacy of the work of land ice, though from the 1870s it was largely his younger brother who came to specialize in this area.

A further achievement of Archibald Geikie lay in introducing to British scientists the discoveries of Powell, Gilbert and Dutton in their surveys in the western United States. Chorley, Dunn and Beckinsale indicate (*op. cit.*, vol 1, 596) that few British geomorphologists were noting these findings in their publications. S.W. Wooldridge in fact suggests that it was not until the 1890s that Davis's papers made current the 'new geomorphology' in this country ('The role and relations of geomorphology', 1948, in *The geographer as scientist*, 1956, 82-3), but the 'new mode' was present in *The scenery of Scotland* (1865). Geikie's American visit in the late 1870s was followed by papers in *Macmillan's Magazine*, included in the collection *Geological sketches at home and abroad* (1882). Prior to this he had been publishing details of the American Geological Surveys in *Nature*. In the third edition of *The scenery of Scotland* (1901) (Preface, vi) he makes the point that had British workers been confronted with the landscapes of the Far West, there would have been no question of their accepting so readily the marine erosion theory. Between the first and second editions of that work, (1865 and 1887), Geikie was able to claim that his formerly controversial views had become part of the common stock of knowledge.

For all the influence of this geomorphological work, Geikie's most monumental achievement is his gigantic *Textbook of geology*, which surveys the whole field of that subject. Yet even in this it is clear that his contemporaries felt that its dynamical, structural and physiographical aspects surpassed those details which form the heart of present day geology. Thus Lapworth contrasts the 'perfunctory, didactic and defensive' stratigraphical section of the text with the dynamical, 'the finest and most attractive portion of the work, filled with the very spirit of Lyell himself', (Review in *Geol. Mag.*, decade II,

vol 10, 42). The concluding words may be taken as symptomatic of where much of Geikie's heart lay: 'In that bourne [the ocean floor] alone can they [the land sediments] find undisturbed repose, and there slowly accumulating in massive beds, they will remain until, in the course of ages, renewed upheaval shall raise them into future land, and thereby enable them once more to pass through a similar cycle of change' (*Textbook of geology*, 2nd ed, 1885, 943). This is not the only anticipation of Davis in Geikie's work. The great man himself later identified in Geikie's discussion of the factors affecting the origin and evolution of scenery in Scotland the fundamental principles of his 'structure, process and stage' model (W.M. Davis, 'Experiments in geographical description', *Scott. Geogr. Mag.*, 1910, 27, 568).

As G.L. Davies points out (*op. cit.*, 356), the Geikies were the last of a line, for the twentieth century has seen a decline in the influence of geomorphology in British university geology departments, and its transfer into departments of geography.

Geikie was a strong advocate of the need to link man and environment. It is not strictly correct to imply, as Preston James seems to (*All possible worlds*, 1972, 257), that Geikie was laying claim to the influence of the physical features on man as part of the field of geology, for Geikie clearly saw this as in the nature of geography. While on occasions deterministic overtones crept in, in defining the nature of the interaction between man and his environment he noted how man 'has changed the face of nature, and how, on the other hand, the conditions of his geographical environment have moulded his progress' ('Geographical evolution', *Proc. R. Geogr. Soc.*, 1879, 1, 423).

The idea of the region lies close to the surface of some of his work. Linton credits him with initiating the threefold and perhaps too slavishly followed distinction of the major regions of Scotland. These Geikie describes (*The scenery of Scotland*, 3rd ed, 1901, 108) as marked by their own 'peculiarities of geological structure and external configuration'. In a chapter entitled 'The physical geography of a region' in *The teaching of geography* (1887) he refers to the 'great provinces or regions of the earth's surface, characterized by particular assemblages of living things... ' (2nd ed, 1908, 180).

Apart from the approbation accorded to his geomorphological contributions by Chorley, Dunn and Beckinsale and G.L. Davies, it may be that Geikie's overall contribution to geography has been given insufficient credit. In the last quarter of the nineteenth century physical geography was vigorously studied but its position in relation to geography and geology was equivocal. Leading geographers were searching for a synthesis between physical and human aspects, a quest which Geikie fundamentally supported. But his location on the scientific periphery of the subject, and some of his incursions into academic debates, could be interpreted as indicating a less than wholehearted sympathy. His main lapse, in the eyes of geographers, may have been his reply to a communication from Keltie, printed in the latter's report to the Royal Geographical Society (1886), in which Geikie does not support the case for the establishment of chairs of geography in universities.

Similarly, at a later date, he rejects Markham's plea for a clean line of division to be drawn between geology and physical geography, regarding the role of the two as 'inextricably interwoven.... Not only is the present the key to the past, but it is equally true that the past is the key to the present.' Any line of division would be an artificial one, drawn purely for the purposes of convenience (see discussion on 'The limits between geology and physical geography', *Geogr. J.*, 1893, 2, 533-4).

In 1879, eight years before the publication of Mackinder's famous paper 'On the scope and methods of geography' (*Proc. R. Geogr. Soc.*, 1887), Geikie stresses geography's role in establishing connections between scattered facts, in ascertaining the interrelationships between different parts of the globe, and 'the function of each in the economy of the whole.... It traces how man, *alike unconsciously and knowingly*, has changed the face of nature, and how, on the other hand, the conditions of his geographical environment have moulded his own progress.' Like Mackinder, Geikie valued the humanizing, bridge-building qualities of geography, bringing to problems 'a central human interest in which these sciences are sometimes apt to be deficient' (Geographical evolution', *Proc. R. Geogr. Soc.*, 1, 1879, 423). The great difference between Mackinder and Geikie was that Mackinder's paper was part of a campaign to establish geography in the universities, whereas Geikie, presumably for political reasons (his own chair of geology at Edinburgh had not been established until 1871), was unsympathetic to this cause.

De Lapparent, the French geologist, attributed much of the success of Geikie's books to the fact that 'the traditional barrenness of geology is always smoothed and adorned by a deep and intense feeling for nature. Nobody has done more than he to associate geological science with the appreciation of scenery' ('Scientific worthies: Sir Archibald Geikie', *Nature*, 1893, 47, 219). Not surprisingly, Geikie was a keen environmentalist and conservationist. When plans were publicized for a tramway up Snowdon, he protested. While he accepted the need to open up 'our mountain solitudes...to the general public', limits had to be set. 'Can nothing be yet done to save this geological sanctuary from the vandalism of the modern company promoter?' (Letter to *Nature*, 1901, 64, 206). At the same time Geikie was haunted by a longstanding, though declining, literary tradition which saw the collecting activities of natural scientists in the countryside in a similar despoiling light, and he must have quoted Wordsworth's lines with some pain:

> Ye may trace him oft
> By scars which his activity has left
> Beside our roads and pathways.

('English science and its literary caricaturists', *Haslemere Natural History Society Science Paper No 6*, 1914, 42.)

Geikie was a 'connector' *par excellence*. Scenery appealed both to his scientific spirit and poetic instincts. He constantly enjoined fellow scientists to develop literary skills and learn foreign languages. He saw a training in scientific methods alone, without the humanizing influence of literature, as narrowing and was an ardent advocate of integrating the 'newer' and the 'older' learning. He

denounced utilitarian views of the value of scientific learning as 'pernicious'. Scientists of this ilk he described as 'not always the most agreeable members of society...apt to be somewhat angular and professional, contributing little that is interesting to general conversation...' ('Science in education', 1898, reprinted in *Landscape in history and other essays*, 1905, 287).

3. INFLUENCE AND SPREAD OF IDEAS

Geikie was always concerned to place things in context, not least himself and his writings. His deep interest in the historical development of his field of study is reflected in his biographies of Forbes, Murchison, Ramsay and Michell, in *The founders of geology* (1897), in shorter papers on Miller, Darwin and others, and in his many obituary notices of British and foreign geologists. He was acutely conscious of the interdependence of the global scientific community, with which he had contacts throughout the length and breadth of Europe, in North America and in Japan. He was revered abroad as the embodiment of British geology. The wide diffusion of his work reflected its innate distinction, his contacts made during travels abroad, his command of foreign languages, and the fact that many of his books were translated into other languages.

Geikie's skills as a communicator were instrumental in making him one of the most influential teachers of the century. His university lectures were delivered in a clear and attractive way, but Strahan notes that although they were elegantly structured, 'in delivery he was not the equal of many a far less able man in rousing enthusiasm in his audience', (*Q.J. Geol. Soc.*, obit.). He was an important innovator in university teaching methods. Despite lack of suitable initial accommodation, he introduced practical laboratory work, including microscopic investigation of thin rock sections, a recently developed petrographical technique. He built up a collection of pictorial representations, helped by his artistic gifts, and of rocks, minerals and fossils.

Most significant of all was his emphasis on the field excursion, including both day trips in the Edinburgh neighbourhood, and weekly trips further afield to areas for which he had equal affection, such as Arran and Speyside. Prior to the local excursions, students were frequently invited to breakfast at the Geikie residence. He noted that the field excursions had social benefits, strengthening the bond between teacher and student. W.A. Herdman, later Professor of Natural Science at Liverpool, was one of Geikie's students, and claimed that the inspiration derived from these excursions led to the endowment of the Herdman Chair of Geology in Liverpool. This gesture gave Geikie the keenest pleasure of his professional career.

Though never a schoolteacher himself, Geikie was passionately interested in the progress of science, including geology and geography, in the schools. He went to great trouble in preparing a handbook (published) for the geological collections at his son's school, Harrow. He wrote articles on topics such as 'The use of Ordnance Survey maps in teaching geography' (1902), and 'Science teaching in public schools' (1913).   But it is for his textbooks that Geikie was known above all.   Prodigious though his efforts were in compiling advanced texts, he found even more demanding the task of preparing his two little Science Primers (in a series edited by T.H. Huxley) on *Physical geography* (1873) and *Geology* (1874).   'Simplicity of treatment, brightness of style, and clearness in exposition were the chief qualities aimed at' (*A long life's work*, 161).

Sound educational principles had been ingrained from early experiences and contacts.   Geikie's first fossil collecting excursion to Burdiehouse Quarry left a sense 'of the enormous advantage which a boy or girl may derive from any pursuit that stimulates the imagination' ('My first geological excursion' in *Geological sketches at home and abroad* (1882, 24). He was also much influenced by Pillans, an educational reformer who was an energetic opponent of rote learning methods.   The concern to infuse meaning and stimulus was a central objective in Geikie's teaching through textbooks.   A reviewer of one of his texts noted that at the root of Geikie's teaching was the principle 'that geography should be a real and not merely a verbal study' (*Scott. Geogr. Mag.*, 1888, 287).

While previous 'physical geography' texts, particularly Mary Somerville's (1848), had achieved popularity, T.H. Huxley (*Physiography*, 1877) and Geikie introduced new dimensions of scholarship and pedagogic insight.   Thus Huxley's text begins evocatively on London Bridge and uses the Thames Basin as a case study.   Geikie's *Geology* starts with the building materials of an ordinary dwelling house, and employs a problem-posing framework, with chapter headings such as 'A quarry and its lessons', and 'How the rocks of the crust tell the history of the earth'.

Geikie's treaching strategies are laid out in his methodological text *The teaching of geography* (1887), described by twentieth-century writers as 'surprisingly modern' (Bramwell, R.D., *Elementary school work, 1900-1925*, 1961, 28), and as anticipating in principle all that is today emphasized as important in 'local studies, field work and active participation in geographical learning generally' (Gopsill, G.H., *The teaching of geography*, 1956, 15).   Geikie regarded geography as crucial in education, for its incomparable breadth provided a means of integrating literature, history and science.

The input into school geography of a truly scientific approach was of great importance in improving a tarnished intellectual image, associated with elementary schooling and its attendant rote learning. Of course the intellectual imbalance between the physical and historical/political aspects of geography created great problems for its development as an academic subject, whose resolution was the achievement of Mackinder and Herbertson.   It must be said that Geikie was more successful in applying his tenets in the physical than in the human field of geography. Certain of the chapters of both *The teaching of geography* and of *An elementary geography of the British Isles* (1888) are tainted by the repetitious use of counties as divisions, and also by an all-embracing concept of the field of 'human geography'. Mackinder was critical of this over-broad sweep, illustrations being proposed of 'many subjects which

even the most grasping geographer would scarcely claim as his'.   While he praised the book's stress on scientific explanation and personal observation, he clearly felt that the methodology was more to do with pedagogy in general, than with geography teaching in particular.

Geikie's concern with birds, beasts, insects, laws of health, exchange by barter, and so on, was related to his immersion in the *Heimatskunde* tradition, but the environmental educationist of today would understand him.   And Geikie was the prototype environmental educationist, seeing local studies as not only stimulating interest and powers of observation and induction, but also as training the young in treating the landscape, physical and cultural, with 'loving care and respect.   The rude and ribald spirit of desecration which cuts names on...worm-eaten woodwork, scribbles over...walls, or breaks off corners from...mouldering carvings, ought to be absolutely repressed' (*The teaching of geography* 2nd ed, 134).

At all levels Geikie had a passionate desire to communicate meaning and a joy in nature.   From his formative years he was a bridger of cultures and was aware before his time that the earth was a shrinking place.   His ability to 'connect' was of truly cosmic proportions.

'In fine, looking back across the long cycles of change through which the land has been shaped into its present form, let us realise that these geographical revolutions are not events wholly of the dim past, but that they are still in progress.   So slow and measured has been their march, that even from the earliest times of human history they seem hardly to have advanced at all. But none the less are they surely and steadily transpiring around us.   In the fall of rain and the flow of rivers, in the bubble of springs and the silence of frost, in the quiet creep of glaciers and the tumultuous rush of ocean waves, in the tremor of the earthquake and the outburst of the volcano, we may recognise the same play of terrestrial forces by which the framework of the continents has been step by step evolved.   In this light the familiar phenomena of our daily experience acquire an historical interest and dignity.   Through them we are enabled to bring the remote past vividly before us, and to look forward hopefully to that great future in which, in the physical not less than in the moral world, man is to be a fellow-worker with God.   ('Geographical evolution', *op. cit.*, 352.)

Bibliography and Sources

1. REFERENCES ON ARCHIBALD GEIKIE

a. Obituary notices
Important obituary notices include the following:
Proc. R. Soc., vol III (1926), xxiv-xxxix
Proc. R. Soc. Edinburgh, vol 45 (1925), 346-61
Nature (1924), vol 114, 758-60.

Q.J. Geol. Soc., vol 81 (1925), lii-lx
Proc. Geol. Assoc., vol 36 (1925), 191-2
Bull. Geol. Soc. Am., vol 36 (1925), 128-9
Z. Vulkanol., vol 9, 149-55
The Times, 12 November 1924

b. Reviews of autobiography, *A long life's work*,
include *Nature*, vol 114 (1924), 114-18. (by A.
Strahan); *Geol. Mag.*, vol 61 (1924), 515-19; *Scott.
Geogr. Mag.*, vol 40 (1924), 249-50 (by H.M. C(adell))

c. Biographical (apart from autobiography, Geikie. A.,
A long life's work (1924))
Geol. Mag., Decade III, vol 7 (1890), 49-51
A. de Lapparent, Scientific Worthies Series, *Nature*,
vol 47 (1893), 217-20 'Account of complimentary dinner
given to Sir Archibald Geikie on his retirement as
Director General of the Geological Survey', *Nature*,
vol 64 (1901), 34-6
P. Macnair and F. Mort (eds) *History of the geological
society of Glasgow, with biographical notices of
prominent members* (Glasgow, 1908). Sir Archibald
Geikie, 254-7

d. Bibliographical
A definitive bibliography is that of E. Cutter, 'Sir
Archibald Geikie: a bibliography', *J. Soc. Bibliogr.
Nat. Hist.*, vol 7 (1974), 1-18
This contains 246 references.

2. SELECTIVE BIBLIOGRAPHY

In view of the comprehensive character of Cutter's bibliography, the following is highly selective, and
includes only important papers or texts on the overlap
region between geology and physical geography, and on
educational topics.

1862 'Geological notes on the Auvergne', in Galton, F.
(ed), *Vacation tourists and notes of travel in
1861*, London, 211-48. Reprinted as 'Among the
volcanoes of central France', in *Geological
sketches at home and abroad*, London, (1882)
86-126
1863 'On the phenomena of the glacial drift of
Scotland', *Trans. Geol. Soc. Glasgow*, vol 1,
1-190
1865 *The scenery of Scotland viewed in connection with
its physical geology*, London, 2nd ed 1883, 3rd ed
1901; 360p.
1866 'Notes for a comparison of the glaciation of the
west of Scotland with that of Arctic Norway',
Proc. R. Soc. Edinburgh, 5, 530-56. Reprinted
as 'The old glaciers of Norway and Scotland', in
Geological sketches at home and abroad, London,
(1882), 127-66
1868 'On modern denudation', *Trans. Geol. Soc.
Glasgow*, vol 3, 153-90
1873 *Physical geography*, Science Primers Series,
Macmillan, London, 2nd ed 1884; 3rd ed 1900;
110p.
1874 *Geology*, Science Primers Series, London, 135p.
1876 *Outlines of field geology*, London, 2nd ed 1879;
3rd ed 1882; 4th ed 1891; 61p; 5th ed 1912;
260p.
1877 *Elementary lessons in physical geography*, London,
363p.

1877 *Questions on Geikie's elementary physical
geography*, London, 101p.
1879 'Geographical evolution', *Proc. R. Geogr. Soc.*,
vol 1, 422-43. Reprinted in *Geological Sketches
at home and abroad*, London, (1882), 312-52
1882 *Geological sketches at home and abroad*, London,
382p. *Textbook of geology*, London, 2nd ed 1885;
3rd ed 1893, 971p; 4th ed 2 vols 1903, 1472p.
1884 'The origin of the scenery of the British
Islands', Royal Institution Lectures, *Nature*,
vol 29, 325, 347-8, 396-7, 419-20, 442-3.
Reprinted in *Landscape in history and other
essays*, London, 130-57
1886 *Classbook of geology*, London, 2nd ed 1890;
3rd ed 1892; 4th ed 1902; 5th ed 1909; 6th ed
1915; 516p.
1887 *The teaching of geography: suggestions regarding
principles and methods for the use of teachers*,
London, 2nd ed 1892, 202p.
1888 *An elementary geography of the British Isles*,
London, 2nd ed 1904, 127p.
1893 'Scenery and the imagination', *Fortn. Rev.*,
vol 53, 547-73
Reprinted as 'Landscape and the imagination' in
Landscape in history and other essays, London,
28-75
1897 *The ancient volcanoes of Great Britain*, London
2 vols
1898 'Types of scenery and their influence on literature', the Romanes Lecture, in *Landscape in
history and other essays*, London, 76-129
'Science in education', Lecture to the students
of Mason University College, Birmingham.
Reprinted in *Landscape in history and other
essays*, London, 282-307
1902 'The use of Ordnance Survey maps in teaching geography', *Geogr. Teach.*, vol 1, 61-7
1904 'Continental elevation and denudation', *Q.J.
Geol. Soc.*, vol 60, lxxx-civ
1905 *Landscape in history and other essays*, London,
352p.
1906 'The history of the geography of Scotland',
Scott. Geogr. Mag., vol 22, 117-34
'The origin of landscape', *Edinburgh Rev.*,
vol 204, 356-80
1908 'The Weald', *Trans. Southeast Union Sci. Soc.
for 1908*, 1-20
1913 'Science teaching in the public schools',
Nature, vol 90, 555-6
1914 'English science and its literary caricaturists
in the 17th and 18th centuries', *Haslemere Nat.
Hist. Soc. Sci. Pap.*, no 6, 45p.
1924 *A long life's work: An autobiography*, London,
426p.

3. UNPUBLISHED SOURCES

Edinburgh University Library. The bulk of the Geikie
archives are under the accession number E60/33
and consist mainly of copies of lectures, geological
notebooks, and correspondence to professional colleagues
at home and abroad.
National Library of Scotland. More isolated pieces
of correspondence, dating from 1873-1908, in volumes
1, 2, 4 and 7 of the catalogue to the archives.
Royal Society. Contains around 200 pieces of corres-

pondence of which the most extensive is with his
friend Sir Joseph Larmor, Senior Secretary of the Royal
Society.
Royal Geographical Society. A few letters indicating
a friendly relationship with J. Scott Keltie, Secretary
of the Society.

*W.E. Marsden is a Senior Lecturer in Education,
University of Liverpool.*

CHRONOLOGICAL TABLE: ARCHIBALD GEIKIE

DATES	LIFE AND CAREER	ACTIVITIES, TRAVEL, FIELDWORK	PUBLICATIONS	CONTEMPORARY EVENTS AND PUBLICATIONS
1835	Born 28 December in Edinburgh			
1842				Opening of Glasgow-Edinburgh railway, the first in Scotland
1845	Entered Edinburgh High School	First visit to Scottish Highlands		
1851		First excursion (Arran) away from Edinburgh for geological study		
1852	Entered Edinburgh University			
1854			First paper on lias of Skye for Royal Physical Society of Edinburgh	Crimean War
1855	Appointed to Geological Survey of Scotland	Field mapping in Lothians, Fife, Ayrshire and Renfrewshire (to 1869)		
1857			First paper on geology of Strath (Skye) to Geological Society of London	
1858			First book: *The story of a boulder*	
1859		Elected to Geological Society of London	First paper to British Association (Aberdeen)	Darwin's *Origin of Species*
1860/61	Lectured at School of Mines, London			
1861	Fellow of Royal Society of Edinburgh Examiner in geology at London University	First trip abroad to study volcanic phenomena of the Auvergne	*Memoir of Edward Forbes* (with G.Wilson); paper with Murchison on 'Highland structures'	
1862			'Geological notes on the Auvergne', in Galton, F. (ed), *Vacation tourists* Geographical map of Scotland (with R. Murchison)	
1863			Paper 'On the phenomena of the glacial drift of Scotland'	
1865	Fellow of Royal Society	Visit to Arctic Norway to study glaciation	*The scenery of Scotland*	

DATES	LIFE AND CAREER	ACTIVITIES, TRAVEL, FIELDWORK	PUBLICATIONS	CONTEMPORARY EVENTS AND PUBLICATIONS
1867	Director of newly formed Scottish branch of the Geological Survey President of Geological Section of British Association (Dundee)		Presidential address 'History of volcanic action in the British islands'	
1868		Visit to the Eifel and Switzerland to study volcanic and glacial phenomena	Paper 'On modern denudation'	
1869		Innsbruck meeting of German and Swiss naturalists	Memoirs of Geological Survey (Ayrshire, etc.) with James Geikie	
1870		Visit to southern Italy to study volcanic phenomena		Franco-Prussian war
1871	Married Alice Pignatel; Appointed first Murchison Professor of Geology at Edinburgh University (to 1882); Again President of Geological Section of British Association (Edinburgh)		Presidential address 'The geology of Edinburgh and its neighbourhood'; Inaugural lecture 'The Scottish school of geology'	Death of Sir Roderick Murchison
1873		Visit to southern Italy	First school text *Physical geography* (in T.H. Huxley's Science Primer Series); 'The Ice Age in Britain' (Manchester Lectures)	
1874			*Geology* (Science Primer Series)	*The Great Ice Age*, James Geikie
1875			*Life of Sir Roderick Murchison;. Outlines of field geology*	
1877			*Elementary lessons in physical geography*	
1879	Macdougall-Brisbane medallist of Royal Society of Edinburgh	Lectures at Lowell Institute, Boston. Visit to volcanic areas of western U.S.A.	'Geology' (*Encyclopaedia Britannica*); 'Geographical evolution' (*Proc. R. Geogr. Soc.*); 'On the old red sandstone of western Europe' (*Trans. R. Soc. Edinburgh*)	

DATES	LIFE AND CAREER	ACTIVITIES, TRAVEL, FIELDWORK	PUBLICATIONS	CONTEMPORARY EVENTS AND PUBLICATIONS
1881	Murchison medallist of Geological Society of London; Member of Athenaeum Club			
1882	Director General of the Geological Survey of Great Britain (to 1901)		'Charles Darwin's work in geology', in *Charles Darwin: Memorial notices reprinted from 'Nature'*; *Geological sketches at home and abroad*; *Textbook of geology*	
1883		Member of Council of the Geological Society of London		
1884			'The origin of the scenery of the British islands' (Lectures to the Royal Institution)	Sends Peach and Horne to re-investigate Highland structures; Public retraction of views on the issue in *Nature*
1885		Member of Council of the Royal Society		3rd International Geological Congress in Berlin
1886			*Class-book of geology*	
1887			*The teaching of geography*	
1888			*An elementary geography of the British Isles* (These formed the first two volumes of Macmillan's Geographical Series, edited by Geikie)	
1889	Foreign Secretary of the Royal Society	Visit to Norway to study metamorphism	'The history of volcanic action during the Tertiary period in the British Isles' (to Royal Society of Edinburgh)	
1891	President of the British Association; Knighthood	Wintered in Paris (1891-2) on sick-leave		
1892	Governor of Harrow School (to 1922)	President of British Association (Edinburgh); Member of Royal Commission on the water supply of London; Member of department committee to enquire into the condition of the Ordnance Survey	Presidential address 'James Hutton's "Theory of the Earth"'; *New geological map of Scotland with descriptive notes*	Introduction by Geological Survey of monographs descriptive of the major divisions of the geological record

DATES	LIFE AND CAREER	ACTIVITIES, TRAVEL, FIELDWORK	PUBLICATIONS	CONTEMPORARY EVENTS AND PUBLICATIONS
1893			'The limits between geology and physical geography'; Geikie chairs joint meeting of sections of the British Association (Nottingham); 'Scenery and the imagination' (*Fortn. Rev.*)	
1894				International Geological Congress in Zürich
1895	Wollaston medallist of the Geological Society of London; Royal medallist of the Royal Society	Cruise round the Faroes; visits to Switzerland and France (the latter for the centenary of the Institut de France)	*Memoir of Sir Andrew Crombie Ramsay*	
1896			'The Tertiary basalt plateaux of north-western Europe', (*Q. J. Geol. Soc.*).	
1897		Lectured at John Hopkins University, Baltimore; 7th International Geological Congress at St Petersburg; followed by 'Grand Tour' of southern Russia, return via Mediterranean and Italy	*The ancient volcanoes of Great Britain; The founders of geology; Geological map of England and Wales with descriptive notes*	
1898			'Types of scenery and their influence on literature' (Romanes Lecture, Oxford)	
1899	President of Geological Section of British Association (Dover)	Visit to Italy; International Geological Congress in Paris; revisited Auvergne	Presidential address 'Geological time'; Editor of vol. iii of Hutton's *Theory of the earth*	
1900			Memoir *The geology of central and western Fife and Kinross*	
1901	Retirement; Hayden Gold Medal of the Academy of Natural Sciences of Philadelphia	Royal Society delegate to International Association of Academies meeting in Paris		Death of Queen Victoria
1902		Wintered (1902-3) with family in Italy	Memoir *The geology of eastern Fife*; 'The use of Ordnance Survey maps in teaching geography', *Geogr. Teach.*	
1903	Secretary of the Royal Society (to 1908)	Member of the Science and Scholarships Committee of the Royal Commission for the Exhibition of 1851	New (4th) and ex-panded edition of *Textbook of geology* (1472 pages)	International Geological Congress in Vienna

DATES	LIFE AND CAREER	ACTIVITIES, TRAVEL, FIELDWORK	PUBLICATIONS	CONTEMPORARY EVENTS AND PUBLICATIONS
1904			*Scottish reminiscences*	
1905	Livingstone Gold Medal of the Royal Scottish Geographical Society		*Landscape in history and other essays*	
1906	President of the Geological Society of London	In Paris to give address to the Alliance Française	'The history of the geography of Scotland' (*Scott. Geogr. Mag.*)	
1907	K.C.B.; Governor of the Imperial College of Science and Technology	In Sweden as Royal Society delegate to the bi-centenary celebrations of the birth of Linnaeus	Editor of Faujas de Saint-Fond's *A journey through England and Scotland to the Hebrides in 1784*	
1908	President of the Royal Society (to 1913) Trustee of the British Museum	Foreign Secretary of the Geological Society of London	Centenary address (Geological Society) 'The state of geology at the time of the foundation of the Geological Society'; 'The Weald' (*Trans. Southeast Union Sci. Soc.*); *Charles Darwin as geologist* (Rede Lecture, Cambridge)	
1910	Death of son, Roderick, in a railway accident; President of the Classical Association	Royal Society delegate to International Association of Academies meeting in Rome	Notes on Seneca's 'Questiones Naturales' in J. Clarke, *Physical science in the time of Nero*	
1912			*The love of nature among the Romans*	
1913	Order of Merit Moved house to Haslemere, Surrey	Visit to Italy to study landscapes among which the Roman poets grew up		
1914		Return to Italy for further pilgrimages	'English science and its literary caricaturists in the 17th and 18th centuries'	Outbreak of the Great War
1915	Death of daughter Elsie, and of brother James, his successor as Professor of Geology at the University of Edinburgh since 1901		'Catallus at Rome', *Q. Rev.*	
1916	Death of wife		'The birds of Shakespeare'	
1917			*Annals of the Royal Society Club*	
1918			*Memoir of John Michell*	

DATES	LIFE AND CAREER	ACTIVITIES, TRAVEL, FIELDWORK	PUBLICATIONS	CONTEMPORARY EVENTS AND PUBLICATIONS
1919	Chairman of Royal Commission of Inquiry into Dublin University (Trinity College, Dublin)			
1924	Died 10 November, aged 88, at Haslemere, Surrey		*A long life's work* (autobiography)	

Other awards and honours received include - Officier de la Légion d'honneur; Foreign Associate, Institut de France; Member of the Council of the British School in Rome; Gold Medal of the Institute of Mining and Metallurgy. Geikie was also a member of the following Academies: Lincei, Rome; Berlin; Vienna; Petrograd; Belgium; Stockholm; Turin; Naples; Munich; Christiania; Göttingen; Philadelphia; New York; National Academy of Sciences of the United States.   He received honorary degrees from the Universities of Oxford; Cambridge; Dublin; Edinburgh; Glasgow; Aberdeen; St Andrews; Durham; Birmingham; Sheffield; Liverpool; Uppsala; Leipzig; Prague; Strasbourg.

James Geikie
1839–1915

W. E. MARSDEN

1. EDUCATION, LIFE AND WORK

A younger brother of Archibald Geikie (*q.v.*), James was the third child of a family of eight. Having retired from business in Edinburgh, his father devoted himself entirely to music. His uncle was Walter Geikie, a well-known Scottish painter, and his maternal grandfather a sea-captain, whose descriptions of voyages opened the minds of the Geikie children to the outside world. A wide range of interest was thus engendered early in their lives.

After some unhappy years at a private school, James Geikie entered Edinburgh High School where, like Archibald, he shone as a classical scholar. In his leisure hours he joined Archibald and friends on nature rambles and fossil collecting expeditions in the Edinburgh neighbourhood. In 1854 James was apprenticed to an Edinburgh printer, but found the confinement and long hours of an office job so trying that it affected his physical and mental health. The appointment of Archibald to the Geological Survey in 1855 was decisive. Leaving his apprenticeship in 1858, James spent some time attending natural history lectures, including geology, at the University, and held temporary posts until a vacancy occurred in the Geological Survey in 1861.

James Geikie's early years were spent in pioneer surveys of the drift deposits. Previously only the solid geology had been mapped. He also joined Archibald in mapping in Ayrshire, and later turned to the Lanark coalfield. His mapping experiences of the lowland drifts stimulated James's interest in glaciation. Visits to Norway in 1865, and with his brother, Peach and Horne in 1868 to the Rhineland and Switzerland were important in helping him to visualize the Europe of the Great Ice Age.

In 1872 he transferred to Kelso to undertake mapping in the Border country. The combined effort of surveying and the writing of his major work *The Great Ice Age* led to an attack of nervous prostration in the winter of 1872-3: nevertheless the book went off to the printers in 1873, and this with a tour of Europe lifted his spirits for a time. But the introspection recurred, as a letter of 1874 reveals: 'Every day teaches a man how really little after all he does know, for the beliefs which yesterday he held with all the tenacity of conviction, are today shaken by doubts, and tomorrow may be cast aside as effete'.

In 1875 things looked up again. He married, was elected a Fellow of the Royal Society, and completed his work on the Borders. He moved north to Perth, and began surveying in the Perth-Dunkeld district. The success of *The Great Ice Age* led to demands for a new edition. Darwin expressed his admiration for the work, and the letters from 'old father Darwin' were regarded as 'very precious' by Geikie. In 1876 he was asked by Ramsay to join him in an investigation of the water supplies of Gibraltar, which later led to a joint paper. In 1877 a second important article 'On the glacial phenomena of the Long Island or Outer Hebrides' appeared (*Q.J. Geol. Soc.*, *1878*). Further articles on glaciation and climatic change and a second major book, *Prehistoric Europe: a geological sketch* (1880) were produced during these last years with the Geological Survey. The work involved brought on further outbreaks of depression.

Archibald Geikie's appointment as Director Gen-

eral of the Geological Survey of Great Britain in 1882
left the Chair of Geology at Edinburgh vacant.   His
younger brother was an obvious contender, but the
thought of moving from the Survey in his early forties
caused James great heart-searching.   Despite con-
sidering it poorly paid, Geikie thought the security
of a university post attractive.   Nevertheless, he
made it clear that he would take the Directorship of
the Scottish Geological Survey, which post Archibald
had also held in a dual capacity, should this be
offered.   In fact it was not.   But the eminence of
his writings was such ample testimonial for appointment to the Chair that no other candidate was even
approached.

In his inaugural address, 'The aims and methods
of geological inquiry' (1882), James Geikie spoke of
the boon of taking over a department possessing a
class museum and teaching aids, handed on from his
brother, and also the burden of a reputation 'which
must bear hard upon me.   He has not only sustained
but increased the fame of what has been termed the
Scottish School of Geology, and I feel that it will
task all my energies to emulate the high standard he
maintained as a teacher.'   Geikie found the effort
of lecture preparation a chore.   He had no assist-
ants and there were certain technical aspects of
geology on which he was less than expert.   He also
missed the camaraderie of his Geological Survey
colleagues, for which he compensated by inviting
members of the Survey to social occasions with
students.

In 1884 Geikie visited Canada and the U.S.A. in
connection with the British Association meeting in
Montreal.   The same year saw the establishment of
the Scottish Geographical Society, with which he was
involved from the start, becoming Honorary Editor in
1888 and later Vice-President and President (1904-10).
From the first publication of the *Scottish Geographical Magazine* he contributed regular articles, usually
on topics in the borderland between physical geography
and geology.   A number of these papers were collected
together, with others, to make up *Fragments of earth
lore: Sketches and addresses, geological and geographical* (1893).

His first decade in the Chair at Edinburgh wit-
nessed the consolidation of his reputation as a
writer.   As was the case with his brother, teaching
responsibilities prompted the appearance of textbooks.   His *Outlines of geology* (1886) proved a success and by 1903 had run to four editions.   His
health broke down again between 1886 and 1888, and
recuperation was sought in holidays in the Canaries
(1887) and in the Engadine and Italy (1888).   In
1889 he was awarded the Macdougall-Brisbane Medal of
the Royal Society of Edinburgh and the Murchison Medal
of the Geological Society of London.   In 1890 and
1892 he was President respectively of the Geological
(Newcastle) and Geographical (Edinburgh) sections of
the British Association, at which he read papers on
'Glacial geology', (*Geol. Mag.*, 1890) and 'The geographical development of coastlines', (*Scott. Geogr.
Mag.*, 1892).   In 1891 he returned to the United
States to give lectures at the Lowell Institute,
Boston.

His main writing commitment from 1892-4 was the
preparation of a new edition of *The Great Ice Age*,
expanded from just over 600 to 850 pages to include
new material accumulated from the growing international literature on the subject.   In the course
of this preparation, Geikie visited the Baltic coasts
to study the glacial deposits.   Once more the extra
work resulted in a setback in health.   In 1894 James
Geikie was made Dean of the Faculty of Science at
Edinburgh, giving him a position of greater influence,
and although he doubted his own capacity as a committee man, his long tenure of the post suggests that
he made a success of it.   He continued to produce
articles for the *Scott. Geogr. Mag.* including 'The
morphology of the earth's surface' (1895), an early
use of the term 'morphology' in the British literature.   His next significant work was his major contribution to the science of 'geomorphology', *Earth
sculpture, or the origin of landforms* (1898).

By the turn of the century, Geikie had passed
his 60th birthday, but continued unabated in his
writing and teaching.   He produced three further
books: *Structural and field geology for students*
(1905), which ran to six editions, the last as late
as 1953; *Mountains: their origin, growth and decay*
(1913), a work much influenced by Suess; and *The
antiquity of man in Europe* (1914), as well as further
articles, again largely for the *Scott. Geogr. Mag.*
He continued to travel, for purposes of study, family
holidays, and recovery from bouts of ill-health.

His international reputation was secure.   The
United States Geological Survey named after him a
mountain in Wyoming (1900).   He was made an honorary
member of the New York Academy of Sciences (1901) and
corresponded regularly with colleagues in America,
such as Chamberlin, who had contributed a section on
North America for the 3rd edition of *The Great Ice
Age* (1894).   He took particular pride in being the
dedicatee of Penck and Bruckner's *Die Alpen im
Eiszeitalter* (1909).   In 1913 he was elected
President of the Royal Society of Edinburgh.   He
retired in June 1914, having given his last lecture
at the age of 76, and on 1 March, 1915, he died
suddenly from a heart attack.

James Geikie seems to have been an altogether
earthier character than his elder brother.   In his
autobiography, H.R. Mill contrasts the 'genial,
hearty, plain-spoken' James with the 'precise and
formal' Archibald.   Hinxman describes James as of
'strong personality and vivid imagination; nervous
and quick-tempered, but by nature genial and warm-
hearted; staunch in friendship and formidable in
controversy...' (review of Newbigin and Flett's
biography of James Geikie, *Scott. Geogr. Mag.*, 1918,
34, 67).   While physically extremely robust in his
days in the Geological Survey, he was mentally less
resilient that his famous brother.   He worried about
work, money and the responsibilities of a large
family, four boys and one girl, which arrived relatively late in his life.

It may be that he felt compelled to strive to
emulate the achievements of Archibald, and this too
proved a strain.   Although by normal standards a
highly successful man, James's letters show periods
of exuberance mixed with brooding and even despair.
They reveal a person on the defensive, which may
explain the acidity of some of his comments about the
scientific fraternity of his time.   He was particu-

larly scornful of English academics. In a letter to
his publisher in 1876 he mentions a new geological
discovery to be included in the 2nd edition of *The
Great Ice Age* which had been hailed by Ramsay with
glee and which, according to James, 'settled the hash'
of his English opponents, one of whom was Boyd Dawkins
of Manchester. The journal *Nature* he described as
a 'soil-pipe' carrying off the 'acrid humours' of a
'pettifogging Cockney clique'. It is intriguing that
his brother Archibald was a regular contributor to
Nature. James in general seems to have seen English
dons in a sinister, imperialist light, providing
enlightenment 'for the improvement of us poor hyper-
borean savages'.

His conspiratorial interpretations of the process
of academic advancement come out clearly in a series
of letters to Patrick Geddes, a close friend though
sixteen years his junior, at a time when both were
being linked with vacant chairs at the University of
Edinburgh. Geikie advised Geddes on the canvassing
strategies he saw as necessary to securing appointment. He wrote of the need to gain access to those
acquainted with 'the backstairs political mysteries
and wirepulling'; to engage in 'persistent begging
and brazen-faced importunity'; to 'dine out' in
Edinburgh, 'uttering dulcet sounds of blandishment'.
He referred also to his brother Archibald as trying
to 'counteract evil influences', presumably acting
against Geddes. The attempts to gain Geddes the
Natural History Chair were unsuccessful and Geikie's
own appointment seems to have caused him a little
embarrassment. As he confessed to Geddes, he wished
he had not asked for so many testimonials. 'Canvassing makes a man do much he would laugh at in
others...it might in future teach me charity.'

A calculating element was present in James's
character as it was in Archibald's. He thus
suggested to his publisher that Charles Darwin's
privately communicated admiration of *The Great Ice
Age* might be incorporated in the 2nd edition to 'help
to expedite the sales', adding: 'You see how com-
mercially I look at things'. In his defence, it may
be said that there is evidence that financial con-
straints caused him worry in the early years of his
marriage.

Perhaps lacking the intellectual span and more
cosmopolitan sophistication of his elder brother,
James none the less had a wide range of interests, and
his literary attachments resulted in a book entitled
*Songs and lyrics by Heinrich Heine and other German
poets* (1887). He was also a keen golfer and cyclist.
Like Archibald, he had a strong vein of humour and
affection for Scottish folklore, and was a raconteur
to match. Unlike him, he seems to have had the
capacity for developing close personal friendships and
working relationships, particularly in his years in
the Geological Survey.

2. *SCIENTIFIC THOUGHT AND CONTRIBUTIONS TO GEOGRAPHY*

Like Archibald, James was in the great line of nineteenth-century earth scientists, achieving international recognition. Both were helped by influential mentors, Archibald by Murchison, and James by
Ramsay, though the impact of Ramsay on James's
thought was more comprehensive than that of Murchison
on Archibald's. Ramsay's great influence was in the
field of glacial geology, where James Geikie's most
authentic contribution to the development of scientific thought lies.

Horne wrote: 'It is difficult for the younger
generation to realize the effect produced by the publication in 1874 of his epoch-making The Great Ice
Age' (obituary, *Scott. Geogr. Mag.*). G.L. Davies
(*The earth in decay*, 1968, 311-12) confirms that the
work of Archibald and James Geikie and Croll ensured
that by the late 1870s 'icebergs and the glacial sub-
mergence were no longer factors of importance in
British geomorphic thought'. *The Great Ice Age*
established James Geikie as the leading protagonist
of the land ice theory, as Archibald was by then
devoting his attentions to other matters. It seems
strange that Chorley, Dunn and Beckinsale, in *The
history of the study of landforms*, vol 1 (1964),
while mentioning an important though small-scale
article of 1868 (p.451), pay no attention to James
Geikie's masterpiece.

Even more than the establishment of the land ice
theory, *The Great Ice Age* introduces the notion of
climatic change in the glacial period. This became
the most controversial aspect of the work. James
refused to accept prevailing ideas that Palaeolithic
deposits were post-glacial, suggesting, on the contrary, that they were inter-glacial. He cour-
ageously advanced the view, drawing on his investigations of the Scottish peat mosses, that there were
several inter-glacial periods. While his proposal
of six has not stood the test of time, his basic
principle was triumphantly vindicated. His ideas
were developed in *Prehistoric Europe* (1881), in which
he definitely backdated the earliest known relics of
man to the climatic ameliorations of the inter-glacial
periods. He also discusses the impact of glacial
fluctuations on sea-level, and shows himself in
advance of his time in the interpretation of raised
beaches.

As a geologist James Geikie was, like Archibald,
an avid fieldworker, with an acute eye for country.
He had great ability as a draughtsman, but not the
detailed grasp of the palaeontological, petrographical and mineralogical foundations of his subject
which would have prevented errors made in the mapping
of the Silurian volcanic rocks of Ayrshire, for
example. His greatness lay in the fields of
structural, physical and 'dynamical' geology. Flett
(1917) considers his most successful work was in the
borderland area between physical geology and
geography.

James Geikie laid emphasis on the interaction
between the two. Geography was prerequisite to the
study of geology which, dealing with 'the operations
of Nature in the past' necessitated a 'clear know--
ledge of the mode in which Nature works at present
(*Nature*, 1882, 27, 44). Similarly, physical geography could not function without the backing of
geology. The physical geographer 'can describe the
actual existing conditions; without the aid of
Geology, he can tell us nothing of their origin and
cause'. (*Scott. Geogr. Mag.*, 1887, 3, 401). He
saw the relationship between the two as analogous to
that between political geography and history. Without
the grasp of an area's history, geographical under-

standings' would hardly surpass those of a commercial traveller, whose geographical studies have been confined to the maps and tables of Bradshaw' (the famous railway timetable), (p.400).

Geikie's geographical commitment was cemented in a long-standing and dedicated association with the Royal Scottish Geographical Society, for which he wrote many of his most important papers, acted as Honorary Editor and later as Vice-President and President. Marion Newbigin, who worked closely with him as one of four successive acting editors, wrote of his capacity for delegation and moral support, while at the same time reading all matters submitted for publication, and contriving 'without any parade of interference, to exert a steady influence upon the nature of the contents', (*Scott. Geogr. Mag.*, 1915, 31, 205). Apart from his major texts, it was the Scottish Geographical Magazine which above all became the vehicle for the diffusion of his ideas.

3. INFLUENCE AND SPREAD OF IDEAS

As Archibald Geikie's successor as Professor of Geology at the University of Edinburgh, James Geikie amply achieved his aim of upholding the reputation of the Scottish School of Geology, which Archibald himself acknowledged on James's death in 1915. With his brother in a sense becoming 'anglicized', spending virtually half his life in England, James was left as the doyen of Scottish geology.

His skills as a university teacher proved to be considerable, though he was almost in middle age when he was made professor. Over-defensive at first, and somewhat resentful of the chore of preparing his 'sermons', which had to cover the whole field of geology, he records his relief and an element of condescension when he realizes he has gone down well with his students. In a letter to Patrick Geddes (1882) he writes: 'I lectured them with the satisfying feeling that they were following me like docile sheep' and, after his initial qualms, feels he has found 'the right groove'.

Flett describes James Geikie as having an 'easygoing, colloquial style'. 'He spoke fast, and covered a very large part of his subject in the course of the one hundred lectures which constituted the work of the winter class...' (*James Geikie: the man and the geologist*, 185). To relieve the students of the drudgery of taking voluminous notes, he prepared duplicated notes on a primitive copying apparatus. He preferred large wall diagrams to lantern slides and, like his brother, laid particular emphasis on field excursions and on establishing a social as well as an intellectual rapport with his students.

Among those who attended James Geikie's classes were Patrick Geddes, A.J. Herbertson, and H.R. Mill. Mill regarded his field excursions as of 'unrivalled supremacy' (*An autobiography*, 1951, 31), while Geddes said that under his 'vivid influence', he nearly mistook geology for his profession (P. Boardman, *Patrick Geddes: maker of the future*, 1944, 16). Both these influential men in the geographical world became close friends of James Geikie.

James Geikie's influence abroad was, if anything, greater than at home. Albrecht Penck, for example, saw himself as a Geikie disciple, and on his death

sent a tribute to Geikie's wife:
> James Geikie belongs to those who have influenced most my scientific evolution. His clear way of seeing things and his reasoning made a convincing impression on me, and though I never listened to one of his lectures, I felt always to be one of his students. He was my master. His *Great Ice Age* showed me the ways to understand the glacial deposits of Central Europe: his *Prehistoric Europe* arose my interest for prehistoric questions: his views on mountains, valleys and lakes gave me the base for my morphological work.
(Newbigin and Flett, *op. cit*, 137)

Though critical of details of Geikie's interpretations of glacial phenomena, W.M. Davis refers to *The Great Ice Age* as the 'standard work' on the subject, and quotes at length from *Earth sculpture* ('Glacial erosion in France, Switzerland and Norway' in *Geographical Essays*, ed. D.W. Johnson, 1954 edition), as does C.A. Cotton, as late as 1942, in *Climatic accidents in landscape-making*.

Flett claims that Geikie's influence as a university teacher came high in his list of services to science, his classes, small though they might be, producing 'almost every year one or more men who subsequently made a name for themselves in science. All over the world, and especially in the British colonies, there are many well-known geologists who can trace the impetus which decided their careers to the lectures delivered by the genial professor in the dingy old Edinburgh class-room at the top of those interminable stairs.' (Newbigin and Flett, *op. cit.*, 207-8).

Unlike Archibald Geikie in the 1880s, James became a strong protagonist in the moves to establish geography as a university subject. In a letter to the wife of Patrick Geddes (1905) he speaks warmly of Geddes's famous Outlook Tower in Edinburgh, seeing it as 'a valuable adjunct to a school of geography which we all hope may ere long flourish in Auld Reekie'. No doubt as part of this campaign, an editorial appeared in the *Scott. Geogr. Mag.*, (1905, 1-4) (possibly written by Newbigin with the support of Geikie) recounting the objectives of the Society, namely, 'to press for the full recognition of the claims of geography as a branch of knowledge'. Apart from its manifest usefulness, it provided 'a genuine insight into the methods and results of modern science', the development of which had been 'the great glory of our own time'; while no other subject afforded 'more beautiful examples of the interaction of organism and environment'. Nor was there a better intermediary between scientific specialists and the general public

Like Archibald, James Geikie ardently supported moves to improve geographical education, though he has left us much less evidence of his views on the subject. In his first article in the *Scott. Geogr. Mag.* (1885), entitled 'The physical features of Scotland', he notes the obligation which teachers should have to the maps of the Ordnance Survey:
> With such admirable cartographical work before them, how long will intelligent teachers continue to tolerate those antiquated monstrosities which so often do duty as wall-maps in their school-rooms?... With a well-drawn and faithful orographical map before

him the school-boy would not only have his labours
lightened, but geography would become one of the
most interesting of studies.   He would see in
this map a recognisable picture of a country, and
not, as at present is too often the case, a kind
of mysterious hieroglyphic designed by the
enemy for his confusion.

James Geikie contributed two texts to Chambers'
'Elementary science manuals' series, *Geology* (1875)
and *Historical geology* (1876).   The technique to be
used for the series was described in a preface:

Instead of a number of abrupt statements being
presented, to be taken on trust and learned, as
has been the usual method in school-teaching;
the subject is made, as far as possible, to
unfold itself gradually, as if the pupil were
discovering the principles for himself...now
acknowledged to be the only profitable method
of acquiring knowledge.

Geikie stressed that his aim was 'rather to indicate
the methods of geological enquiry, than to present
the learner with a tedious summary of the results'.

It must be admitted that, as in some sections
of Archibald Geikie's school texts, there would
seem to us today a distinct gap between laudable
intention and achievement.   In James Geikie's
books, the pedagogic device used was to number in
order the 97 paragraphs in the book, then ask
questions on each paragraph.   The publishers
claimed this made self-instruction possible.   In
fact, paragraph one was a definition of geology,
and question one asked 'What is geology?'.   The
information was, however, connected, explanatory,
and of scientific worth, a distinct advance on the
listed morass of information which constituted the
'capes and bays' approach.   At this time there
was no shortage of ideas we would still regard as
progressive and appropriate, but the means to apply
them in practice, including the technological
variety of aids which we now enjoy, was not
available.

James Geikie's most important contribution to
the cause of geographical education was probably
his indirect one, in controlling the destinies
for thirty years of the *Scottish Geographical
Magazine*, which under his aegis maintained a
successful blend of articles on world-wide
exploration, scientific research, and geographical
education, including papers by such notable
pioneers as Scott Keltie, Mackinder, Herbertson,
H.R. Mill, Patrick Geddes and W.M. Davis.

Bibliography and Sources

1. REFERENCES ON JAMES GEIKIE

a. Obituaries
Scott. Geogr. Mag., vol 31 (1915), 202-5, *Q.J. Geol.
Soc.*, vol 72 (1916), liii-lv; *Geol. Mag.*, Decade vi,

vol 2 (1915), 192; *Nature*, vol 95 (1915), 40-1;
Glasgow Herald, 3 March 1915; *The Times*, 3 March 1915.

b. Biographical
Newbigin, M.I. and Flett, J.S. *James Geikie: the man
and the geologist*, Edinburgh (1917), 227p. (This
biography has provided much of the basic material
for the preceding account.)
'Eminent living geologists series', *Geo. Mag.*,
Decade v, vol 10, (1913), 240-8
Macnair P. and Mort F. (eds), *History of the Geolo-
gical Society of Glasgow*, with biographical
notices of prominent members (Glasgow, 1908)
'James Geikie', 270-2
Horne, J., 'The influence of James Geikie's researches
on the development of glacial geology', *Proc. R.
Soc. Edinburgh*, vol 36, (1915), 1-25
Hinxman, L.W., 'James Geikie: his life and work'
(review of Newbigin and Flett, *op. cit.*, 1917),
Scott. Geog. Mag., vol 34, (1918), 66-8

c. Bibliographical
Newbigin, M.I. and Flett J.S. *op. cit.*, (1917),
213-19
Geol. Mag., *op. cit.*, (1913), 245-8
Horne J., *op. cit.* (1915), 20-5

2. SELECTIVE BIBLIOGRAPHY
In view of the comprehensive bibliographies cited
above, the following is selective, and concentrates
on important texts and papers in the overlap area
between geology and physical geography.
1867 'On the buried forests and peat mosses of Scotland, and the changes of climate they indicate',
Trans. R. Soc. Edinburgh, vol 24, 363-84
1868 'On denudation in Scotland since glacial times,'
Trans. Geol. Soc. Glasgow, vol 3, 54-74
1871 'On changes of climate during the glacial epoch',
Geol. Mag., vol 8, 545-53; vol 9, (1872), 23-31,
61-9, 105-11, 164-70, 215-22, 254-65
1872 'On the glacial phenomena of the Long Island or
Outer Hebrides', *Q. J. Geol. Soc.*, vol 29, 532-45
1874 *The Great Ice Age and its relation to the antiquity of Man*, London, 2nd ed, 1894, 575p.
1875 *Geology* (Chambers 'Elementary science manuals'),
London, 96p.
1880 'On the glacial phenomena of the Long Island or
Outer Hebrides' (2nd paper) *Q. J. Geol. Soc.*,
vol 34, 819-67
1880 *Pre-historic Europe: a geological sketch*,
London, 592p.
1882 'The aims and methods of geological inquiry'
(Inaugural lecture), *Nature*, vol 27, 44-6, 64-7
1885 'The physical features of Scotland', *Scott. Geogr.
Mag.*, vol 1, 26-41*
1886 *Outlines of geology*, London, 2nd ed, 1888; 3rd ed
1896; 4th ed 1903
1886 'Mountains: their origin, growth and decay',
Scott. Geogr. Mag., vol 2, 145-62*
1886 'The geographical evolution of Europe', *Scott.
Geogr. Mag.*, vol 2, 193-207*

1887 'Geography and geology', *Scott. Geogr. Mag.*,
    vol 3, 398-407*
1890 'The evolution of climate', *Scott. Geogr. Mag.*,
    vol 6, 59-78*
    'Glacial geology' (Presidential address to
    British Association, Section C), *Geol. Mag.*,
    Decade iii, vol 6, 461-77*
1892 'On the glacial succession in Europe', *Trans.
    R. Soc. Edinburgh*, vol 37, 127-49*
1892 'Geographical development of coastlines' (Presi-
    dential address to British Association, Section
    E), *Scott. Geogr. Mag.*, vol 8, 457-79*
1893 *Fragments of earth lore: sketches and addresses,
    geological and geographical,* Edinburgh, 428p.
    This contains, *inter alia,* the papers marked*
    above.
1895 'The morphology of the earth's surface', *Scott. Geog.
    Mag.*, vol 11, 56-67
1895 'Classification of European glacial deposits',
    J. Geol., vol 3, 241-69*
1897 'The last Great Baltic glacier', *J. Geol.*,
    vol 5, 325-39
1898 *Earth sculpture, or the origin of landforms,*
    London; new ed 1902; 2nd ed 1909, 320p.
1901 'Mountains', *Scott. Geogr. Mag.*, vol 17, 449-60;
    18 (1902), 76-84
1905 *Structural and field geology for students,* Edin-
    burgh, 2nd ed 1908; 3rd ed 1912; 4th ed 1920;
    5th ed (revised R. Campbell and R.M. Craig) 1940;
    6th ed 1953, 435p.
1906 'From the Ice Age to the present', *Scott. Geogr.
    Mag.*, vol 22, 449-63
1909 'Calabrian earthquakes', *Scott. Geogr. Mag.*, vol
    25, 113-26
1911 'The architecture and origin of the Alps', *Scott.
    Geogr. Mag.*, vol 27, 393-417
1913 *Mountains: their origin, growth and decay,* Edin-
    burgh, 311p.
1914 *The antiquity of man in Europe,* Edinburgh, 317p.

3. UNPUBLISHED SOURCES
Edinburgh University Library. A small amount of
material, including notes of lectures and a letter to
his brother, Sir Archibald Geikie.
National Library of Scotland. A considerable number
of letters written between 1871 and 1906, in volumes
2, 4 and 7 of the catalogue to the archives. The
great bulk consists of correspondence with Patrick
Geddes.
Royal Geographical Society. A number of letters.
Other material includes a collection of correspondence
from Geikie to his publisher, Edward Stanford, in
connection with the preparation of a second edition
of *The Great Ice Age,* covering the period 1876-8.

*W.E. Marsden is a Senior Lecturer in Education, Uni-
versity of Liverpool.*

DATES	LIFE AND CAREER	ACTIVITIES, TRAVEL, FIELDWORK	PUBLICATIONS	CONTEMPORARY EVENTS AND PUBLICATIONS
1839	Born 23 August in Edinburgh			
1850	Entered Edinburgh High School			
1854	Apprenticed to printer – later attended natural history classes at Edinburgh University			Crimean War
1861	Joined the Geological Survey	Early work mapping drift deposits		
1864–72		Surveys of coalfields of Ayrshire and Lanarkshire, in early years with Archibald Geikie		
1865		Visited Norway to study glacial features		Archibald Geikie, *The scenery of Scotland* published
1866			First publications in *Q. J. Geol. Soc.* and *Geol. Mag.*	
1867			First glacial paper 'On the buried forests and peat mosses of Scotland', *Trans. R. Soc. Edinburgh*	
1868		Rhine journey with Survey colleagues, Archibald Geikie, Peach and Horne; visited Eifel and glaciated areas of Alps	'On denudation in Scotland since glacial times', *Trans. Geol. Soc. Glasgow*	
1871			1871-2 Papers on the Carboniferous formations of the west of Scotland; 'On changes of climate during the glacial epoch', *Geol. Mag.*	Franco-Prussian War 1870-1
1872	Transferred to surveying 'border country'- centred on Kelso			
1873		Long tour of Europe: Italy, France, Switzerland, Austria, Rhineland, Holland	'The antiquity of Man in Britain', *Geol. Mag.*	
1874			*The Great Ice Age and its relation to the antiquity of Man*	
1875	Marriage Elected Fellow of the Royal Society		*Geology* ('Elementary science manual')	

DATES	LIFE AND CAREER	ACTIVITIES, TRAVEL, FIELDWORK	PUBLICATIONS	CONTEMPORARY EVENTS AND PUBLICATIONS
1876	Transferred to surveying Perthshire and Forfar; Moved house to Perth	Visit to Gibraltar with Ramsay to investigate water supply situation; Visited North Africa	*Historical geology* ('Elementary science manual')	
1877	LI.D. St Andrews			
1878		Visit to Faroes to study glacial phenomena		
1880			'On the glacial phenomena of the Long Island or Outer Hebrides', *Q.J. Geol. Soc.* *Pre-historic Europe: a geological sketch*	
1881		Visit to Iceland		
1882	Succeeded Archibald Geikie as Professor of Geology at the University of Edinburgh		'The aims and method of geological enquiry' (Inaugural lecture), *Nature*	
1884	Began active work with the Royal Scottish Geographical Society	Visit to North America – New York, Chicago, St Paul, Winnipeg, Toronto, Montreal (British Association), Philadelphia, Boston		Foundation of Royal Scottish Geographical Society
1885			'The physical features of Scotland', *Scott. Geogr. Mag.*	
1886			'Mountains: their origin, growth and decay'; 'The geographical evolution of Europe', both in *Scott. Geogr. Mag.; Outlines of geology*	
1887		Visit to Canary Islands to recuperate from period of ill health	'Geography and geology' *Scott. Geogr. Mag.; Songs and lyrics by Heinrich Heine and other German poets*	
1888	Honorary Editor of *Scottish Geographical Magazine*	Family visit to the Engadine and Italy		
1889	Macdougall-Brisbane Medal of the Royal Society of Edinburgh; Murchison Medal of the Geological Society of London			

DATES	LIFE AND CAREER	ACTIVITIES, TRAVEL, FIELDWORK	PUBLICATIONS	CONTEMPORARY EVENTS AND PUBLICATIONS
1890	President of the Geological Section of the British Association (Newcastle)		'The evolution of climate', *Scott. Geogr. Mag.*; 'Glacial geology' (Presidential address to British Association), *Geol. Mag.*	
1891		Visit to United States to give lectures at the Lowell Institute, Boston		
1892	President of the Geographical Section of the British Association (Edinburgh)		'On the glacial succession in Europe', *Trans. R. Soc. Edinburgh* 'Supposed causes of the glacial period'; 'The late Andrew Crombie Ramsay'; both in *Trans. Edinburgh Geol. Soc.* 'Geographical development of coastlines', *Scott. Geogr. Mag.* (Presidential address to the British Association)	
1893			*Fragments of earth lore: sketches and addresses, geological and geographical*	
1894	Dean of the Science Faculty, Edinburgh University	Visit to Baltic lands to study glacial deposits of north Germany and Denmark		
1895			'Scottish interglacial beds', *Geol. Mag.*; 'The morphology of the earth's surface', *Scott. Geogr. Mag.*	
1898			*Earth sculpture, or the origin of landforms*	
1900		Visit to Pyrenees		
1901	Honorary Member of New York Academy of Sciences		'Mountain structure and its origin', *Int. Mon.*; 'Mountains', *Scott. Geogr. Mag.*	
1903		Visit to Norway		
1904–10	President of Royal Scottish Geographical Society			

DATES	LIFE AND CAREER	ACTIVITIES, TRAVEL, FIELDWORK	PUBLICATIONS	CONTEMPORARY EVENTS AND PUBLICATIONS
1905			*Structural and field geology for students*	
1906			'From the Ice Age to the present', *Scott. Geogr. Mag.*	
1907			'Old Scottish volcanoes', *Scott. Geogr. Mag.*	
1908		Family holiday in Portugal		
1909			'Calabrian earthquakes', *Scott. Geogr. Mag.*	
1911		Holiday in Switzerland	'The architecture and origin of the Alps', *Scott. Geogr. Mag.*	
1912			'The deeps of the Pacific Ocean and their origin', *Scott. Geogr. Mag.*	
1913	President of the Royal Society of Edinburgh		*Mountains: their origin, growth and decay*	
1914	Retirement from the Chair of Geology at Edinburgh University		*The antiquity of Man in Europe*	
1915	Died 1 March, aged 76, in Edinburgh			

Edmund William Gilbert

1900–1973

T. W. FREEMAN

For 44 years from 1923, E.W. Gilbert worked as a geographer in the Universities of London, Reading and Oxford. He was a writer of distinction and a man of wide culture, having a deep love for the English countryside where much of his leisure time was spent. He cared almost instinctively for the welfare of others, as much of his printed work shows, and made interesting contributions to applied geography, though throughout his career his interest in an historical approach to geography was apparent.

1. EDUCATION, LIFE AND WORK

Son of Rev. R.H. Gilbert, M.A., Rector of Hemsworth, Yorkshire, E.W. Gilbert was brought up in a comfortable and somewhat conservative Anglican household. Though located in a mining area, Hemsworth is surrounded by rich farmland and like his father, who managed to find time to hunt twice a week, 'Billy' Gilbert enjoyed rural pleasures including horse riding. He was always a dedicated worker but valued his recreations such as the theatre, wandering on foot through the countryside, photography, and travelling abroad as well as in Britain and Ireland. He was an acute observer wherever he went and his interest in towns, especially those of some charm and historical character such as university cities and seaside resorts, led to some notable writing, especially his *Brighton: old ocean's bauble*, published in 1954. This book, however, he regarded as partly a social history (as indeed it was) and he chose to signify its special character by writing his name differently on the title page as Edmund W. rather than his usual

E.W. Gilbert. Like many other geographers he thought that people too conscious of orthodoxy might make a 'pure' geography so extreme that it would be dull and restrictive in writing and lecturing.

Only by chance did Gilbert become a geographer. He went to St Peter's School, York as a classical Exhibitioner, with the obvious expectation that he would become either a clergyman or a schoolmaster after graduating in Oxford or Cambridge, more probably the former as his father had entered the ministry after graduating from Jesus College, Oxford. He went to Oxford in 1919 as an Exhibitioner of Hertford College to read History, in which he graduated with Second Class Honours in 1922. It was a hard time, for his father died just before he sat his final examinations and the shock probably prevented his achieving a much-coveted 'first'. He was advised to acquire a qualification as a teacher and attended the summer school in Oxford given biennially by the staff of the School of Geography with visiting lecturers. From 1922-3 he followed the Diploma in Geography course, and in 1924 he presented a thesis on the area around Pontefract and Doncaster, in which his home was situated. For the examination papers and this thesis he was given the Diploma, with 'exceptional merit' for the 'geographical description' (as the thesis was then called) and also the Herbertson Prize. It was a basis for his first paper, on Pontefract, in the *Geographical Teacher*, in 1925.

In October 1923 Gilbert became Junior Lecturer in Geography at Bedford College for Women, a constituent college of the University of London, where he taught cartography and surveying, historical geography and the history of geographical exploration. Uncertain whether geography was really his life's work, he considered a career in law, sat in 1924 for the Criminal Law and Procedure examination, and was given a first class mark. In 1925 Gilbert heard a lecture on China by Professor P.M. Roxby of Liverpool and from that time never doubted that he should devote his life to geography. The dynamic personality, inspired oratory, belief in peace and progress characteristic of Roxby's fascinating mind captivated many young, as well as more mature, people. After three years in London, which he found uncongenial, Gilbert went to the University of Reading in 1926 as Lecturer in Historical Geography. In this year, too, he married a fellow student, Barbara Flux Dundas, who assisted him in his work for 47 years.

In 1928 Gilbert was awarded the B.Litt. degree at Oxford for work on historical geography. At that time geography was developing in the British universities, but though there were some Honours graduates from Cambridge, London, Liverpool and Aberystwyth, the Oxford Diploma was also regarded as a fine qualification especially with a good degree in another subject. Undoubtedly Gilbert was influenced throughout his life by the historical training given in Oxford, and the lack of books and other materials of a geographical character came to him as a shock in the Diploma course. There was in his eyes a great opportunity for geographical writing.

Gilbert stayed for ten years at Reading, working with one colleague, A.A. Miller, who was by training a geologist. An Honours course was established in 1926. Like other academics, Gilbert was associated with colleges for the training of primary teachers, six of which were connected with Reading University. As an extension lecturer for the Oxford Delegacy for Extra-mural studies he gave courses of lectures in various towns. During his time at Reading Gilbert's main publication was *The exploration of western America; An historical geography* issued in 1933 by the Cambridge University Press, an extension of his earlier work for the B.Litt. degree. Gilbert, however, never went to America. Much more widely read and greatly appreciated was his article in the *Scottish Geographical Magazine* in 1932, entitled 'What is historical geography?' which among several possible definitions included 'the regional geography of the past'. This view was apparent in his contribution dealing with 'The human geography of Roman Britain', in *An historical geography of England before 1800*, edited by H.C. Darby (1936).

The return to Oxford in 1936 as Research Lecturer in Human Geography was financed by the Rockefeller Foundation's support of the Social Studies Research Committee. One book promoted by this committee was the *Survey of social services in the Oxford district* (1938), to which Gilbert contributed a chapter on the area's geography. At this time three of the four major strands that were to characterize the work of Gilbert's middle and later years were apparent. Articles on the Balearic Islands, then little known, appeared in 1934 and 1935, on inland and seaside holiday resorts in 1939 and on practical regionalism in 1939. Significantly, the first three were published in the *Scottish Geographical Magazine*, then at the height of its career, with editors who chose

articles for their interest and merit rather than for their place of (Scottish) origin. The studies of the Balearics led to Gilbert's work as editor and contributor, from 1941-5, on the Admiralty Handbooks of Spain and Portugal (4 vols). The work on the health resorts led to the book on Brighton of 1954, and the study of regionalism in England and Wales, on which the first article appeared in the *Geographical Journal*, was of considerable interest after the war when it became clear that some radical reforms were needed. But so much discussion followed that reorganization was achieved only in 1974, almost thirty years later. Gilbert was much concerned with local planning, and his relatively long article on 'The industrialization of Oxford' was a classical study of its time. Later, in 1958 and 1959, he visited the universities of Marburg, Göttingen, Heidelberg and Tübingen which, with Oxford and Cambridge, were discussed in his monograph, *The university town in England and West Germany* (1961).

A final interest, biographical study, was first revealed in the publication of a long obituary of Sir Halford John Mackinder (1861-1947) in the *Geographical Journal* (vol 110, 1948, 94-9). Never was better material provided, for Mackinder asked Gilbert to visit him and told him the story of his life, and never was material better used. Later publications by Gilbert on Mackinder seemed to some to exaggerate his academic strength but the development of critical biographical studies had its ultimate fruit in the publication of *British pioneers in geography* (1972), only a short time before his final illness. He died at his home in Appleton, Berkshire, in 1973, having retired from his Professorship of Geography in 1967.

2. GEOGRAPHICAL THOUGHT

a. The published work

Many British geographers have written far more than Gilbert, but he cared more for the quality than the quantity of published work. His main contributions were made through articles meticulously written and then carefully revised. As the years advanced, his writing acquired increasing elegance and wit. Like many Englishmen with a background of ancient cathedral and university towns, he was troubled by the effects of industrialization on the amenities and appearance of many fine old towns, and notably of Oxford. Unhappily he never made the comparable study of Cambridge that he had in mind, though he often said that the growth of Cambridge through the impact of its industrial expansion had been more fortunate, or at least less unfortunate, than that of Oxford. This he ascribed, in 1961, to the planning of Cambridge by the County and not by the city itself, as in Oxford.

Throughout his career he was deeply conscious of history, not as something in the past but as something living on into the present. Everything in his formative years was conducive to such a view. Always he loved the Church of England with its historic buildings, its age-honoured liturgy, its association with the ancient universities. In both the English countryside and its towns there was every indication of a history living on into the twentieth century: what he feared was the possibility, even, at times, it seemed the certainty, that a time-hallowed heritage would be squandered and submerged by 'development'. And his education, at a traditionally minded English boarding school and in Oxford, strengthened his strong sense of history. The fortunate encounter, at first merely at a lecture but later in friendship, with P.M. Roxby showed Gilbert that history and geography could always be in close association. This view owed much to the influence of French geographers, most notably Vidal de la Blache, on British academics during the 1920s and 1930s. What mattered, in his view, was that the environmental influence on human enterprise at various periods of history, including the present as well as the past, should be fully appreciated.

In his first book, *The exploration of western America*, he quotes with approval Roxby's statement that 'the real object of historical geography' was 'the reconstruction of the physical setting of the stage in the different phases of development' (p.26). He goes on to say that he was attempting to do this for western America 'before civilised man had begun to alter the face of the country with the scars of mines, towns and railways'. The book is not, however, the work by which he will be most remembered, for although it is interesting to read, the subject was vast and lacks the inspiration that he could have acquired only by living knowledge through fieldwork. Far more successful was his study of Roman Britain in *An historical geography of England before A.D. 1800*, for there he was not only using excellent printed sources, including maps, but was also writing of a land that he knew well. That the Romans had an eye for the physical geography of the lands they conquered has never been doubted, and this Gilbert used for his most readable and scholarly study.

When Gilbert began teaching in Bedford College, London, he followed the pattern of the time by giving lectures on the geographical influences on humanity throughout prehistoric and historic times, and also on geographical discovery from the time of the Egyptians and Phoenicians to modern efforts to reach the poles. Such courses were so broad that inevitably they were regarded by some as superficial. Gilbert was interested in explorers as individuals and used some of their findings in his book on western America. The same interest was seen in his article on 'Richard Ford and his *Handbook for travellers in Spain*' (*Geographical Journal*) - written in recognition of the centenary of the publication of Ford's guidebook by John Murray. Gilbert's article is illustrated, with characteristic taste, by reproductions of five of Ford's watercolours. Earlier reference has been made to the interest of Gilbert in the Balearic Islands (virtually unknown to tourists in the 1930s) and in Spain, which developed initally from holiday journeys. Though this gave him the background for substantial contributions to the Admiralty Handbooks on Spain published during the 1939-45 war, he never contemplated further work on the Iberian peninsula, nor did he consider further work on explorers, though he gave constant encouragement to others who were interested, especially writers for the Hakluyt Society. His interest in personalities made him turn to the study of geographers and their work, especially of Halford John Mackinder (1861-1947) and Andrew John Herbertson (1865-1915), both of whom taught in the Oxford School

of Geography, as well as to Vaughan Cornish (1862-1948), who worked outside all universities but after long and varied enterprise in writing geography, acquired fame as an advocate of the preservation of the countryside. Several of Gilbert's most interesting papers, plus some new material, appear in his *British pioneers in geography*, (1972). It was fortunate that he encouraged others to follow lines of enquiry that appealed to him. In some ways he was pragmatic and practical, understanding how much could be done, and therefore happy to find people who cared to continue the work.

b. *Geography and other studies*

Gilbert was influenced throughout his career by the regional approach of the Oxford school established by H.J. Mackinder and A.J. Herbertson. He recognized the fundamental importance of the physical basis of geography and in many of his writings, notably his paper on Oxford published in 1947, made most effective use of such material. But his own predilections were humanistic. He had none of the contempt for other subjects of study, which has unfortunately been seen in some geographers, and his respect for historical studies was fundamental. He was also fascinated by the influence of landscape on novelists, and in 1960 he wrote that although the geographer often speaks of the 'personality' of a region, the novelist is probably better able to show it. No geographer, he said, reading the novels of Thomas Hardy or Arnold Bennett is wasting time. His 1960 paper on 'The idea of the region' was given as the Herbertson Memorial Lecture to the Geographical Association.

Gilbert was possessed of a strong social conscience and this was abundantly shown in his series of papers on administrative divisions of Britain, with possible regional entities that would have some form of governmental organization, though not necessarily local parliaments. When this work began, many geographers had failed to study administrative boundaries, possibly because there appeared to be so little hope of ever redefining them. The general tendency was to take them for granted. They became of great interest during the 1939-45 war and afterwards, when many essential services were organized on a regional basis and a radical reform of administrative boundaries was promised. Gilbert welcomed the opportunity given to geographers to help in the redefinition of such boundaries.

Some of the work of Gilbert was concerned with the contribution that could be made to other specialisms. The 1972 volume of *British pioneers (op.cit.)* includes an essay on 'Victorian pioneers in medical geography', originally prepared for the I.G.U. Commission on Medical Geography at Washington in 1952. Although Gilbert was well aware of the possibilities of current research on medical geography, this paper is, in effect, a contribution to applied geography of an historical character. Far more importance was attached to his work on towns, on which his studies were eagerly read by those interested in planning. The 1952 paper on the industrialization of Oxford, for example, was reprinted for distribution to all its subscribers by the Oxford Preservation Trust. Maybe this paper revealed a situation beyond redemption, for there was no hope of ever removing the vast motor industry that had arisen almost by an accident of economic history. Yet the continuing watchfulness and care for the preservation of the amenity of Oxford became a national as well as a local concern. Like the study of Brighton it was a contribution to planning, all the more effective because it did not say how future development should be controlled but drew attention to the geographical basis, both physical and human, on which future planning should be carried out. In the works of a revered pioneer planner, Sir Patrick Geddes, the need was for 'survey before action'.

From his training and from his temperamental makeup, it was not surprising that Gilbert's main contacts apart from geographers should be with historians, students of English literature, planners and others such as medical men who had a natural concern for the welfare of humanity.   Nor was it surprising that he venerated the pioneer work of Vaughan Cornish on national parks and the preservation of the lovely countryside of England. That was his heritage and he saw every reason to preserve it for future generations.

c. *The world view*

Gilbert was a young geographer at a time when one aim of teaching in universities was to give students a broad vision of the whole world. At the same time every effort was made to train geographers to see what was before them in any landscape, and a crucial element in the teaching was the field excursion, both in urban and rural areas. It was not unusual that in his early years as a geographer Gilbert should be concerned both with the exploration of western America, obviously a broad study and perhaps too ambitious, and with the area around his Yorkshire home, for like many students before and after him he chose an area he knew well for detailed geographical analysis.   Herbertson wrote on world climatic regions, Mackinder on world political geography, yet both cared for their immediate surroundings and hoped to build up a more satisfying regional synthesis from detailed work to be done by many people over many years.   But Gilbert shared with them the view that geography is concerned with regions, and deplored the denigration of regional work that came after the 1939-45 war.   The controversy over the rival merits of regional and systematic geography seemed to him ridiculous.

A phrase that acquired great popularity in Britain, 'unity in diversity', with its correlative, 'diversity in unity', appeared to him to represent a useful guide for a geographer.   He gave considerable encouragement to those who wished to work in social geography and at the end of the 1939-45 war was anxious that British geographers should contribute to the development of what were then colonies, as he argued with R.W. Steel in 'Social geography and its place in colonial studies'.   Argument on such issues as possibilism and determinism never stirred Gilbert to publication of a polemic character: he cared more about writing geography.   But he was troubled by the tendency of some geographers to drift 'farther and farther away from lay comprehension' and to break up the subject into minute specialisms of so esoteric a character that they might well be absorbed by other disciplines such as geology, history or economics.

In his view this tendency had been carefully avoided by German and Austrian geographers who, however specialized, were still masters of every aspect of the geography of some geographical region.   Although many would disagree, he quoted with approval Hettner's dictum that 'he who does not understand regional geography is no true geographer'.   He regretted that some of his generation, having come into geography from another subject, made little attempt to think about the essential purpose of geography.

Essentially, in Gilbert's view, geography was one of the humanities.   In his later years he found it difficult to understand the enthusiasm for quantification and he was horrified by the idea of 'political involvement', though generally tolerant of the varied political views of his friends.   He wanted geography to give a liberal education that would provide persons with intellectual integrity, trained to reason and reflect on the conditions of mankind in relation to environment, capable of assessing human situations, and likely to be good servants of the state and even of the Church into which he was born and in which he died.   He feared professionalism, which seemed to mean the acceptance of geography as merely self-perpetuating, so that specialists existed to train other specialists, and regarded that -- and in some cases that alone -- as success.   Gilbert felt that while teachers are necessary, and must be trained by other and more mature teachers, geographers should go out into a world beyond classrooms and lecture theatres.   And as his professional career spanned a period of great expansion in geography he saw much of this happen, for in Britain geographers have gone into a widening range of occupations since the 1939-45 war. Not the least of their contribution has been through planning, for geographers have been markedly successful students in many of the postgraduate courses provided for planners.   He followed the progress of planning with interest.

3. *THE INFLUENCE OF HIS WORK*

To some geographers it seemed that Gilbert was steeped in an outlook that belonged to an earlier generation, and that his work on past geographers, such as Mackinder, Herbertson and Roxby, was marked by excessive hero-worship.   In fact no man was more conscious than Gilbert of the breadth of opportunity offered by geography and it was this element in the work of H.J. Mackinder, studied in his 'Seven lamps of geography', (*Geography*, vol 36, 1951, 21-43), that led to his veneration.   That Mackinder never worked out more than a few of his ideas, due to his political and administrative work, mattered little: he was, in modern terminology, an 'ideas man'.   For Gilbert geography must have breadth, and this he found more among French and German geographers than among the British. He was somewhat reserved in his attitude to American geographers.

Nevertheless, the influence of Gilbert has been considerable.   With a very few other people in Britain he did much to stimulate interest in the history of geography, including the writing of studies of individual geographers.   He did much to develop study of the historical geography of England and Wales, not only in the Roman period but also by his work on towns, which showed a deep appreciation of the geography of the Victorian period, until recently much neglected. He showed convincingly that a geographer could contribute much of significance to planning. And to many he showed that geography could be written with elegance, charm and wit. No man wrote more carefully, and he was equally self-critical of his lectures, many of which, on favourite themes, were remembered afterwards as *tours de force*.

Regrettably, perhaps, Gilbert attended few international conferences, as he much preferred the more intimate atmosphere of smaller gatherings such as the meetings of Section E (Geography) of the British Association. He was a staunch supporter of the Royal Geographical Society and served on its Council from 1948-51, 1954-6 and 1959-62. He was also, from 1945, a member of its Publications Committee, which effectively broadened the range of its articles while maintaining the Society's traditional interest in exploration. For his contributions to urban geography he was awarded the Murchison Grant of the Royal Geographical Society in 1967 and earlier, in 1958, he was made an Honorary member of the Gesellschaft für Erdkunde, Berlin, for studies on the regional geography of England and for researches in social geography. He was also a member of the Council of the Hakluyt Society from 1936-41, 1946-50 and 1952-6.

One source of great satisfaction to Gilbert was contact with scholars in other subjects, and he greatly appreciated the company at Hertford College, of which he was a Fellow from 1953, and where, after his retirement, he was still welcome as the College's first Professorial Fellow. Similarly he enjoyed his three years of service, from 1954-7, as a trustee at the Oxford Preservation Trust. A retirement offering was *Urbanization and its problems* (eds R.P. Beckinsale and J.M. Houston, Oxford, 1968), a collection of sixteen essays by colleagues, former students and friends. He appreciated the kindly interest of many non-geographers in his work, not only in London and Oxford, but also in the village of Appleton where his retirement was spent. Illness supervened before he could carry out all the plans he had made for writing, but it was fortunate that he was able to complete his book on *British pioneers in geography* by 1972.

Of few writers could one say more sincerely that in him there was a touch of the artist. His large circle of friends included many of his former students at Reading and Oxford, with many of whom he frequently exchanged letters.

Bibliography and Sources

1. REFERENCES ON E.W. GILBERT
Obituaries include those in *The Times*, 5 and 18 October 1973 and in *Geogr. J.*, vol 140, 176-7 (1974) and *Geography*, vol 59, 68 (1974). There is a bibliography of his works in *Urbanization and its problems*, ed R.P. Beckinsale and J.P. Houston (1968).

2. SELECTIVE AND THEMATIC BIBLIOGRAPHY
Gilbert's first book, *The exploration of western America 1800-1850, An historical geography*, Cambridge (1933) 234p. developed from his B. Litt. thesis and his only other contributions on this theme were two early papers in *Scott. Geogr. Mag.*, vol 45 (1929), 144-54 and vol 47 (1931), 19-28. His main work was on four themes.

a. Urban and regional problems
1925 'Pontefract', *Geogr. Teach.*, vol 13, 130-5
1934 'Reading: its position and growth', *Trans. South-east Union Sci. Soc. for 1934*, 81-90
1939 'The growth of inland and seaside resorts in England', *Scott. Geogr. Mag.*, vol 55, 16-35
1939 'Practical regionalism in England and Wales', *Geogr. J.*, vol 94, 29-44
1944 'The doctrine of an axial belt of industry in England' (with J.N.L. Baker), *Geogr. J.*, vol 103, 49-72
1947 'The industrialization of Oxford', *Geogr. J.*, vol 109, 1-25
1948 'The boundaries of local government areas', *Geogr. J.*, vol 111, 172-206
1954 *Brighton: old ocean's bauble*, London, 275p.
1958 'Pioneer maps of health and disease in England', *Geogr. J.*, vol 124, 172-83
1960 'The idea of the region', *Geography*, vol 45, 157-75
1961 *The university town in England and West Germany* (monograph), Chicago, 74p.
1965 'The holiday industry and seaside towns in England and Wales', *Festschrift Leopold G. Scheidl*, vol 1, 235-47

b. Iberian peninsula and islands
1934 'The human geography of Mallorca', *Scott. Geog. Mag.*, vol 50, 129-47
1936 'Influence of the British occupation on the human geography of Menorca', *Scott. Geogr. Mag.*, vol 52, 375-90
1945 'Richard Ford and his *Handbook for Travellers in Spain*', *Geogr. J.*, vol 106, 144-51
Gilbert was also the editor and part author of four volumes in the Admiralty Intelligence Division's Geographical Handbooks series on Spain and Portugal: vol 1, *The Peninsula* (1941) 264p.; vol 2, *Portugal* (1942) 450p.; vol 3, *Spain* (1944) 680p. and vol 4, *The Atlantic islands* (1945), 371p.

c. Methodology and historical geography
1932 'What is historical geography?', *Scott. Geogr. Mag.*, vol 48, 129-36
1936 'The human geography of Roman Britain', in Darby, H.C. (ed), *An historical geography of England*, Cambridge, 30-87
1945 'Social geography and its place in colonial studies', (with R.W. Steel) *Geogr. J.*, vol 106, 118-31
1951 'Geography and regionalism', in Taylor, G. (ed), *Geography in the twentieth century*, London and New York, 345-71

d. Biobibliographical studies
1951 'Seven lamps of geography: an appreciation of

the teaching of Sir Halford J. Mackinder' (with bibliography), *Geography*, vol 36, 21-43

1961 *Sir Halford Mackinder, 1861-1947 - an appreciation of his life and work*, Mackinder Centenary Lecture delivered to London School of Economics and Political Science, 32p.

1965 *Vaughan Cornish and the advancement of knowledge relating to the beauty of scenery in town and country*, Oxford Preservation Trust, 21p.

1965 'Andrew John Herbertson: 1865-1915. An appreciation of his life and work', *Geography*, vol 50, 313-31

1965 'Andrew John Herbertson, 1865-1915', *Geogr. J.*, vol 131, 516-19

1972 *British pioneers in geography*, Newton Abbot, 271p. This work includes the essays listed here under (d) with Richard Ford (b) above and some other material, notably an essay on P.M. Roxby.

T.W. Freeman is Emeritus Professor of Geography, University of Manchester. This study is based on his personal knowledge and on assistance from Mrs Barbara Gilbert.

CHRONOLOGICAL TABLE: EDMUND WILLIAM GILBERT

DATES	LIFE AND CAREER	ACTIVITIES, TRAVEL, FIELDWORK	PUBLICATIONS	CONTEMPORARY EVENTS AND PUBLICATIONS
1900	Born 16 October at Hemsworth, Yorkshire			
1914	Entered St Peter's School, York			
1919	Student at Hertford College, Oxford			
1922	B.A. with second class Honours in History			
1923	Junior Lecturer, Bedford College, London University			
1924	Oxford Diploma in Geography with distinction, Herbertson Prizeman			
1925			First article on Pontefract	
1926	Lecturer in Historical Geography, Reading University			
1928	B.Litt. degree, Oxford	Completed thesis on western America		
1930				World economic crisis
1932			Article on 'What is historical geography?'	
1933		First travels in Spain and Balearic Islands	First book, *The exploration of western America*	
1934			First article on Balearics	
1936	Research Lecturer in Historical Geography, Oxford University; Council of Hakluyt Society (to 1941)		Published 'The human geography of Roman Britain' in volume edited by H.C. Darby	H.C. Darby (ed), *An historical geography of England before A.D. 1800*
1939	Lecturer in Geography, Hertford College, Oxford		Articles on regionalism and on inland and seaside health resorts	Second World War begins
1940		Admiralty work; part author of four volumes on Spain and the Atlantic islands		
1943	Reader in Human Geography, Oxford University			

DATES	LIFE AND CAREER	ACTIVITIES, TRAVEL, FIELDWORK	PUBLICATIONS	CONTEMPORARY EVENTS AND PUBLICATIONS
1945	Publications Committee, Royal Geographical Society		Articles on Richard Ford and (with R.W. Steel) on 'Social geography in colonial studies'	In Great Britain, increased interest in town and country planning and in revision of local government boundaries
1946	Council, Hakluyt Society (to 1950)			
1947			In *Geogr. J.*, 'The industrialization of Oxford', and obituary of H.J. Mackinder	
1948	Council, Royal Geographical Society (to 1951)			
1949			Article on Brighton, *Geogr. J.*	
1952	Council, Hakluyt Society (to 1956)			
1953	Professor of Geography and Fellow of Hertford College, Oxford			
1954	Inaugural lecture, 'Geography as a humane study', 12 November; Council, R.G.S. (to 1956); Trustee, Oxford Preservation Trust (to 1957)		Published *Brighton: old ocean's bauble*	
1958		Visit to German university towns	'Pioneer maps of health and disease'	
1959	Council, R.G.S. (to 1952)			
1960		Visit to Stockholm for I.G.U. Congress		International Geographical Union Congress, Norden (Stockholm)
1961			Monograph on *The university town in England and West Germany*	
1964				International Geographical Union Congress, London
1965			Articles on A.J. Herbertson and Vaughan Cornish	

DATES	LIFE AND CAREER	ACTIVITIES, TRAVEL, FIELDWORK	PUBLICATIONS	CONTEMPORARY EVENTS AND PUBLICATIONS
1967	Retired from Oxford Chair of Geography; Awarded Murchison Grant by R.G.S.			
1968		Presentation of *Urbanization and its problems* as a *Festschrift*		
1972			Published *British pioneers in geography*	
1973	Died 2 October at Appleton, Abingdon			

Johannes Gabriel Granö

1882–1956

OLAVI GRANÖ

The career of the Finnish geographer, J.G. Granö, spanned the first half of this century, a time when his subject was seeking its own place in the universities. Granö was intimately associated with geography's attempts to establish its own identity. However, his view of how geography as an independent discipline should participate in the learning process and the development of science differed in principle from the general opinions of his day. His work as an expeditionist, archaeologist, geologist, geomorphologist and cartographer in Central Asia in the first two decades of this century took the form of traditional research and he is respected as one of the classical representatives of the traditional approach. His concept of geography as a scientific explanation of the environment based on data perceived through the senses was novel, although it did not spark off any immediate response. The notion of reciprocity between mind and landscape had not at that time received serious attention from geographers.

Yet in this respect, according to Torsten Hägerstrand, it could be said that Granö was half a century ahead of his time (Vaart, 1975). The modern 'perceptual' school of geography has directed its attention towards the same problems as Granö without realizing the existence of the practical solutions he achieved. Using ideas from the psychology of perception, he tried to combine an examination of the earth's surface with the exact methods of a science by carrying out studies of the real physical environment. Consequently, he developed his own 'environmental' approach to geography, starting from the immediate and distant surroundings or landscape and proceeding to geographical regions. The operational realization of this approach meant that new cartographical methods had to be created. In this he was a recognized pioneer.

1. EDUCATION, LIFE AND WORK

J.G. Granö was born on March 13, 1882, in Lapua, a village in western Finland, where his father, Johannes Granö, a local farmer's son, was the clergyman. His mother came from an old family of merchants and clergymen. Granö had an unusual childhood, for the family moved after a few years to Omsk, in western Siberia, where Johannes Granö served as chaplain to Finnish deportees during the years 1885-91, travelling widely in the southern parts of the country. In 1892 the family returned to Finland, where Granö's father was assigned to a parish just south of the Arctic circle not far from the Swedish border.

Granö made his first acquaintance with geography during his first years as a student at the University of Helsinki. In the autumn of 1901, when attending the lectures of Professor Johan Evert Rosberg (1864-1932), who had just been appointed to the first chair in geography, Granö decided to adopt geography as his subject rather than botany, which had been his main interest. Johann Axel Palmén (1845-1919), the professor of zoology, and Wilhelm Ramsay (1865-1928), the professor of geology, also exerted a strong influence on Granö. Palmén was at this time active in the Geographical Society of Finland, founded in 1888, and Ramsay was working on problems of glacial geology, with which Granö's early research was to be closely connected.

Geography had only recently been accepted as a

university discipline in Finland, due mainly to the efforts of Ragnar Hult (1857-99), a well-known plant sociologist, who in 1890 became the first docent in geography at Helsinki University. Examinations in geography as a university subject were instituted in 1892. Hult became conversant with trends in European geography and searched for a concept of the region. He studied the works of Alexander von Humboldt in detail and was influenced also by Ferdinand von Richthofen. Rosberg, Hult's senior student, continued his mentor's work and through him Granö first became interested in geology and geomorphology. Granö also became concerned about problems of regional geography and these were to dominate his interests some twenty years later. As a young student, Granö had an opportunity to become acquainted with Siberia, for his father was once more appointed chaplain to Finnish deportees in 1902. It was probably not until 1905 that Granö began to give his attention to the problems that were later to become his main interest in his work in Asia. It was at this time that he joined his father on a journey from Omsk to the Altai mountains and as a result was stimulated to further study, since the glacial features he observed were not mentioned in the available literature covering the area.

In order to finance his expeditions Granö accepted a grant from the Finno-Ugric Society awarded him on the initiative of the chairman, Professor Otto Donner. Under the terms of the grant he was asked to look for, photograph and describe old grave mounds with Turkic inscriptions, which were then of particular interest to Finnish archaeologists and scholars of Finno-Ugric languages. Granö spent the summers and autumns of 1906, 1907 and 1909 making long and arduous journeys to collect extensive detailed material about the grave mounds. The results of his work were published in three papers in the journals of the Finno-Ugric Society

On these journeys Granö also did a considerable amount of his own work. The many observations that he made in the formerly glaciated areas of the mountains of southern Siberia and northern Mongolia led eventually to a new concept of the Quaternary history of the regions, for Granö disproved the prevailing view that there had been no extensive Quaternary glaciation in Central Asia.

The first journey, in 1906, lasted five months. The route of the expedition lay south across the Sayan range, along the valley of the Yenisei to the lake district of northern Mongolia. It then turned back north along the old trail through 'Dzungaria's passage'. During the two subsequent expeditions, Granö's interest was directed principally to northwestern Mongolia but he also travelled widely in the Russian Altai, as it was the easiest way to Mongolia. On his third journey he explored as far as Urga, as the capital of Mongolia was then called.

Using the material he had collected on these expeditions, Granö proved his theory about glaciation in the area was correct. He published his results in his doctoral thesis 'Beiträge zur Kenntnis der Eiszeit in nordwestlichen Mongolei' in 1910. At this time, Granö's research reached a point at which further fieldwork was essential. A grant for three years (1913-15) gave him the opportunity to study the Altai mountains, the most suitable area for his investigations. His six years of research in the Central Asian mountains led to the publication of two large works, 'Les formes du relief dans l'Altai russe et leur génese' in 1917 and 'Das Formengebäude des nordöstlichen Altai (which appeared much later, in 1945), and of a number of shorter papers, the last of which was read at the 1949 International Geographical Congress in Lisbon. These papers reveal a great change in Granö's approach to research. The presence of Quaternary glaciations and their maximum extent were documented by means of detailed geological analysis of material, which also enabled him to demonstrate the existence of different glaciations. Granö had become increasingly interested in glacial morphology and, eventually, he dealt with the geomorphology of the whole mountain area.

With the outbreak of the First World War, Granö's opportunities for further work in Asia came to an end. In 1919, after being lecturer at the University of Helsinki since 1908, he accepted the first chair in geography at the University of Tartu in Estonia, at that time a foreign country to him. He was to stay there for four years. The University of Tartu (Dorpat, in Swedish and German, Derpt or Yuryev in Russian) had been founded in 1632 at the time of Swedish rule but it was only in 1919, after Estonia has achieved independence, that it became the national university. Geography was recognized as a distinct discipline and was a separate department in the Faculty of Natural Sciences. The creation of a new discipline and a new university department led Granö to place even greater emphasis on geography as an independent, monistic science with its own object of study the geographical areas of man's perceptual environment. Thus, in 1922, he produced the first definition of geographically homogeneous areas, a definition based on considerable material from Estonia, presented in maps and charts. In a number of subsequent publications in Estonian and Finnish he developed his ideas of geography as an environmental science centering on man's perceptual abilities. While in Estonia, Granö organized intensive local and regional research on both the countryside and the towns, and Estonia provided considerable support for his work.

In the spring of 1923, Granö returned to Helsinki as professor of geography. The following summer he pursued further his theory of the perceptual environment, the inductive treatment of the subject and cartographic representation. Granö published his theoretical studies in this field in 'Reine Geographie' which appeared in German in 1929 and in Finnish in 1930. He also became secretary of the Geographical Society and editor of the third edition of the *Atlas of Finland*. As editor of the *Atlas* he was able to implement many of his ideas on regional studies, and compiled single-handed the first large-scale map showing the distribution of vegetation and population in Finland. It was an achievement that has not since been equalled.

In 1926 Granö was offered the first chair of geography at the University of Turku, which meant that he had once more to establish a department. In addition to the considerable burden of work that all these positions entailed, he continued to apply his ideas to the regional geography of Finland

and Mongolia and, together with his students, to the settlement geography of Finland.

After the Second World War he returned to Helsinki in 1945 to his old position, which he occupied for a further five years. There he initiated and led work on the *Handbook on the geography of Finland* from 1948 to 1952, writing several chapters for the book. He also initiated work on the fourth edition of the *Atlas of Finland,* published in 1960. He was one of the first Finnish geographers to suggest that geography could be a valuable tool in regional planning.

The organizing of research expeditions, the setting up of two new departments of geography, and his work in different academic societies all provide ample evidence of Granö's administrative ability. He was rector of the University of Turku 1832-4 and its chancellor from 1945-55. He had considerable linguistic talents -- in addition to Finnish, his mother tongue, he was also fluent in Swedish, Russian, German and Estonian and had a good knowledge of French and the Tatar language of Central Asia. In 1942 the Government asked him to be ambassador to Sweden although he had no experience as a diplomat and had taken no part in politics, but he refused the offer on the grounds of his academic work.

2. *SCIENTIFIC IDEAS AND GEOGRAPHICAL THOUGHT*

a. *Geomorphology of Central Asia*

Penck and Brückner's work on glaciation in the Alps was used by Granö as a model for his first published studies of glacial geology and morphology. However, he could not apply their methods or use their results from the Alps for his work on the mountains of Central Asia as there are no terminal moraines at the mouths of valleys in the Altai like those in the Alps. Granö therefore had to develop a new method. His careful observations led him to understand the importance of tributary valleys. He realized that river action had continued in these when the main valley was occupied by a glacier, and as the glacier moved down the valley from the high mountains, it dammed the rivers still flowing in the tributary valleys below. As a result, the valleys became filled with material from the glacier as well as from the rivers and quickly reached the final stage of their development. When the glacier later melted, the rivers eroded the deposits in the tributary valleys, forming terraces. During subsequent glaciation the same sequence of events was repeated with the result that two separate terraces were formed at different levels. Terraces of this kind may also be formed in rising mountain regions or when river action changes. However, it may be assumed that the terraces are of glacial origin only in those instances where a connection can be proved between the evidence of river erosion on the upper surfaces of the terraces in the tributary valleys and the features of glacial erosion in the main valley.

Granö's study of the Quarternary geology of the Altai mountains was based on these characteristics. He was able to distinguish two separate glaciations.

He also found indications of two older glaciations but could not determine their extent. Granö was therefore able to prove in his first publications that the theories put forward by the pioneers of glacial research, Gebler (1833) and Kropotkin (1873), and supported by the observations of later scholars (e.g. Sobolev 1896, Saposnikov 1901) concerning the existence of extensive glaciation in Central Asia were correct.

Studies in glacial geology had been closely connected with studies in glacial morphology. The main results achieved in this field had demonstrated the differences in glacial erosion at different altitudes. Granö was able to establish differences between the erosion of corrie glaciers in the higher mountains and the influence of plateau glaciation lower down. The latter preserved the peneplain surface of the lower mountains, below which the glaciers remoulded the former river valleys to create U-shaped valleys. These results were not new in themselves but ice action in relation to existing morphological features had never been so clearly explained.

As a result of Granö's studies in the Altai, a new interpretation was put forward for the formation of these mountains. He had reached the conclusion that the summits of the mountains constituted a late Tertiary raised peneplain, making them younger than had been assumed (e.g. Obruchev, Suess and Machatschek). In his early papers on the morphogenetic interpretation of the mountains, Granö followed Davis's idea of cyclical change to a large extent. Later, however, he made radical changes and used only certain principles from Davis's theory. Grano's later theory, which posed a continuous uplift of the earth's crust, had more in common with the well-known theory of Walter Penck than with Davis's ideas.

b. *Philosophy and methodology of geography and practical applications*

While he was at Tartu, Granö had become interested in the trends in geography that had developed in many centres of geographical research. In his new orientation he was, in his own words, influenced by the 'monists' of geography, above all Hettner (his works after 1903), but also Löffler (1891), Banse (1912), Passarge (1912), Berg (1915), Vidal de la Blache (1913) and del Villar (1915). Granö's new view of geography appeared for the first time in a paper dated 1919 in Tartu. In it he adopted an extremely critical attitude to the general view of the Geographical Society of Finland at the time which had been given publicity by the well-known geologist J.J. Sederholm (*Fennia*, 1912), who referred to C.F. Close (Arden-Close) (*Geogr. J.*, 1911) and others. Sederholm maintained that geography was the sum of different sciences and not a science in itself. General (systematic) geography was, Granö held, an essential part of geography, but there was no justification for regarding geography as a general science for studying the earth. From the ruins of the older, wider concept of geography, a new type of general geography - landscape geography (*Landschaftskunde*) - had to be

created.   In addition, a new approach, which he
called 'regional science', was advocated to replace
the old regional geography (*Länderkunde*), which did
not satisfy the requirements of a science.

Granö favoured the idea of a 'pure geography',
rooted in early nineteenth century *Reine Geographie*.
This recommended a classification, through the study
of natural features, of the earth's surface into
natural regions which have more permanence than
political units.   This discernment of such natural
units gave geography status as an independent science,
rather than being a study ancillary to political
science and history.

Granö shared the conviction, expressed by many
others (e.g. Herbertson, Passarge, Brunhes, Vidal de
la Blache, Berg, Semenov-Tian-Shanski), that the study
of chorological distribution in itself was not
enough.   Instead, it was held that the essential
objects of geographical study are areal units conceived in some way or other as complexes, regardless
of whether such units are termed geographical
entities, regions, landscapes, environments or
milieux.   Many attempts to study these have shown,
however, that in approaching them using scientific
methods one is merely led back to the old general
geography.   For this reason, certain geographers
(e.g. Banse and Younghusband) held that regional
geography had to be raised to the level of an art.
Granö also was of the opinion that regional studies,
carried out by specialists from systematic disciplines, lead to a situation where too little
attention is given to the coexistence and interrelation of phenomena characterizing a region.
Basing his thinking on psychology (principally
Hellpach), he felt that it was possible to consider
the region as a holistic entity only as an impression
of the environment.   Nevertheless he opposed the
view that the scientific period of geography was
drawing to a close and a new artistic period was
approaching.   'As surely as there is an environment,
it is an object of science.   Geography will learn
sooner or later to treat it as such in a scientific
manner ('Reine Geographie', p.6).

For Granö is was a question of the interrelation
between man and his environment, though not of any
causal or possibilistic interaction.   Rather it
was a matter of the reciprocity of man's mind and his
surroundings.   Granö proceeded to formulate a
concept of man's environment as a complex of
phenomena and objects directly perceived by the
senses, of which a unitary impression can be
formed.   Man's environment, for Granö, predominantly consisted of visual, acoustic, olfactory
and tactile sensations which should be considered as
a totality.   Granö made a distinction between the
immediate surroundings, in which we can make
observations with all our senses, and the distant
environment, where our perceptions are realized
mainly through our sight.   This, and only this, he
called landscape.   If we endeavour to treat the
properties of the environment, said Granö, we
have to destroy the barriers that the systematic
sciences have built by dividing the geographical
whole into several different areas of study.   He
conducted his analysis in an exceptional manner, dis-
tinguishing on the one hand between the different

senses, and noting on the other that the properties of the environment appear as either topological
phenomena (in space) or chronological phenomena (in
occurrence).   In both cases they can indicate type,
in which case they are qualitative, or they can
amount, in which case they are quantitative.

Perception in geography meant for Granö understanding the way in which man transforms and structures information from the 'real' physical world
principally as sense data on the spot, not as an
'after-image' or memory.   His attempt to remove
such 'after-images' from geographical studies and to
draw the boundary with sociology and psychology,
derived from his aim that everyone should have a
common 'objective' perception of the real environment, instead of giving priority to their own personal values, attitudes and aesthetic emotions when
reporting the results of their studies or even trans-
mitting geographical data in general.   Obviously
Granö considered the perceptual environment to be
the basis for acquiring our entire *Weltbild*.   He
clearly felt that only through a measurable conceptual description common to all (here he used
symbolic formulae for landscapes) was it possible to
teach, inform and communicate whether it be a
question of observation, experience or knowledge.
It should be borne in mind in this context that at
that time geography was a university discipline with
the single task, in accordance with the ideal of the
unity of teaching and research, of describing and
interpreting the world, and not of helping to
replan it through applied geography.

The surface of the earth, in Granö's view, consisted of a number of different environments bound
by the senses of the person perceiving them.   He
had therefore to consider the relationship between
perceptual experience and external objects, and to
find a way for insights from psychology and the
natural sciences to be integrated within the dis-
scipline of geography.   Such an idea had not been
considered by geographers at that time but certain
psychologists and philosophers had given thought
to the problem.   The main problem for Granö was
to place man's 'subjective' perceptual environment in a real 'objective' environment.   The
environment that encircles man everywhere and continually, both his immediate surroundings and the
landscape, and which moves with the individual observer, has to be fixed in space to the surface of
the earth, to areal units as homogeneous complexes.
These Granö called not landscapes but geographical
regions.

The task of delimiting regions thus defined was
a process of definition rather than discovery: the
regional units were not given but had to be determined arbitrarily to serve the purpose of study.
In drawing up his regional divisions Granö developed
further Passarge's cartographic 'overlay' techniques
of 1908.   In this way he was able to define core
areas and transition zones.   The size of the regions
might vary considerably, depending on the extent
to which he wished either to generalize or go into
detail.   The regions corresponding to landscape
Granö termed tracts, which he grouped together to
form the larger areas of districts and provinces.
These are shown in the *Atlas of Finland 1925*.

Geographical methodology was not for Granö, as for so many geographers before and after him, just conjecture. The goal of his theoretical discussions was always the application of methods to practical research carried out in the real environment. In 'Reine Geographie' these applications are shown in two ways: firstly, in the description and mapping of landscape-scale regions in Estonia; and, secondly, in the illustration of his methods of studying the immediate surroundings because all sensory impressions (not only visual ones) are taken into account. Granö prepared in minute detail a series of cartographic plans of all those sensory impressions and the corresponding localities, 'sites' or 'stows', on the small area of Valosaari in the lake region of Finland. He did not develop or apply this method, which was perhaps his most original one, anywhere else. In the *Atlas of Finland* (1925), 'Geographische Gebiete Finnlands' (1931), and the *Handbook on the geography of Finland* (1952), he did discuss landscapes and corresponding regions. There is also a comprehensive description of the landscapes and regions of Mongolia (1938 and 1941) based on his material from Central Asia. Granö's general philosophical views are contained in his 'Geographische Ganzheiten' (1935).

Granö made his debut as a literary writer in the field of regional geography in 1921, the year in which his book on the Altai appeared in both Finnish and Swedish. It bore the evocative sub-title 'Seen and experienced during years of travel'.

c. Settlement and urban geography
The first object of Granö's studies was the settlement geography of Finnish colonies in Siberia. An important research programme in urban geography was begun under his leadership in Tartu, of which the purpose was to chart the internal patterns of the town, (diurnal traffic flow and pedestrian movement), and urban influence (described in *Terra* 1923). At the beginning of the 1920s, Granö was thus studying inner differentiation in an urban environment and the problems of urban hinterlands. Urban studies as well as studies of rural settlement continued to be carried out in Finland by his students (published in part in *Fennia* from 1930 onwards). Granö himself published a general survey, *Settlement of Finland*, in 1932 and 1952. He also compiled the parts of the *Atlas of Finland* and 'Geographische Gebiete' which dealt with the distribution of population and settlement.

3. INFLUENCE AND SPREAD OF IDEAS
Granö's earliest expeditions to Central Asia resulted in a considerable collection of archaeological and linguistic material. The Danish philologist V. Thomsen had succeeded in interpreting the inscriptions on ancient graves as Old Turkic. Granö's exact notes helped to increase philological knowledge (see Aalto's survey, 1971) and his travel diaries also contained valuable information. It was on the basis of these that a posthumous work was published in 1958. While on his travels Granö accumulated considerable collections of botanical, zoological and geological materials. These are now in the collec-

tions of the universties of Helsinki and Turku.

Granö's first studies in the Altai led many authorities on the Altai at that time, Obruchev and several others (e.g. Saposnikov 1912, Resnichenko 1912, 1914, Jakovlev 1916 and later Bublichenko 1937) to devote more attention to the glaciogenetic origin of the mountains alongside the tectonic and river erosion theories. However, they did not entirely agree with Granö's view. His studies of the Altai glaciation had a certain influence on other glacio-geological studies of mountain ranges (see Merzbacher 1911, Allix 1913, and others).

Granö's new synthetic interpretation of the origin of the Altai range (1917) spread and influenced others in the field (see e.g. reviews by Allix 1923 and Fickeler 1925). It has continued to be highly regarded in surveys of the Altai, although his last major study on the subject, written in German and published in 1945, did not achieve the same status, despite the appearance of a number of notable reviews (Ahlmann, *Ymer* and Cressey, *Geogr. Rev.*). Granö's works have been widely consulted in the Soviet Union as can be seen, for example, from surveys by Obruchev (1926, 1939, 1947), Berg, (1930, 1952), Suslov (1947), Strelkov (1969), and from Gerasimov and Serebryannyj's obituary. The present author was able to ascertain this in the course of a visit to Novosibirsk in 1971 to study a major geological research programme being carried out by the Siberian section of the Soviet Academy of Sciences. As a permanent memorial to Granö's studies of the Altai, a glacier has been named after him. The glacier, on the eastern slopes of the Mongolian Altai, is 8 km long and is situated between glaciers named after two other Altai scholars, Potanin and Krylov.

Granö was an enthusiastic teacher and organizer of university teaching, holding posts at three universities. At Helsinki he was the first assistant in geography and taught courses in cartography as early as 1902, only ten years after geography had achieved the status of an examination subject there. Geography was then a very popular subject and he had many students. His courses and demonstrations in cartography were much more comprehensive than in corresponding departments in many other countries. He attached particular importance to the mathematical base of geographical studies (by this he meant map projections). In Estonia he established his own school of thought, followers of which moved to different parts of the world following the Second World War. Of these the most outstanding was perhaps Edgar Kant, Granö's pupil and close colleague, who moved to Sweden in 1944. Kant won considerable repute as a representative of the Lund school (see e.g. James, *All possible worlds*). In Turku Granö founded his own school. In Helsinki, however, during his last years as a university teacher, his influence was not so marked. It is interesting that, of all the works published by his pupils, not one deals with mountain morphology, and only a few with geomorphology in general. Granö himself wrote a number of works on geomorphological questions, particularly while he was in Estonia, but also following his return to Finland. His lectures were predominantly concerned with physical geography. In his pupils' publications, however, urban studies

and settlement geography constituted the focus of
primary interest.

His ideas of philosophy and methodology of
geography spread rapidly, not only throughout Finland
and Estonia but also to Central Europe: they pene-
trated only sporadically, however, to other parts
of the world.   At the International Geographical
Congresses held in Warsaw and Amsterdam in 1934 and
1938 Granö's work sparked off lively discussions.
His work and ideas were received most favourably in
German *Landschaftskunde* circles, but even there hardly
anybody accepted unreservedly his view that man's perceptual environment should be the basis for geographical
studies.   His view was considered to be too much
influenced by psychology.   With the exception of
the 'perceptual environment', however, the approach
to landscape as synthetic object of study and its
cartographic analysis, were widely approved (see
statements by A. Penck 1928, Hassinger 1930,
Passarge 1931, Lautensach 1933 and 1938, Schrepfer
1934, Burger 1935, Winkler 1938).   But there were
also scholars who took a critical attitude to his
work in other respects.   Among these were
representatives of the *Landschaftskunde* school (such
as Jaeger 1930, and Waibel 1933, and later Lehmann
1950, and Paffen 1953) as well as of other schools
(most critical were Ahlmann in Sweden, and Leiviskä
in Finland).   His influence on the development of
geographical theory in the 1930s was, however,
obvious (see Schmithusen's detailed survey).   The
use of his method of regional division spread to some
extent beyond Finland and Estonia as far afield as
Hungary and Switzerland (cf. Mészáros, Schaffner,
Capt, Windler).

Granö's ideas spread much less in English-
speaking areas, with the exception of his contribution to the *Atlas of Finland*, which attracted
attention everywhere.   It is understandable that
his ideas met with a sympathetic response on the
part of the Sauer-Berkeley school (e.g. Broek).
His works were also reviewed favourably in English
(e.g. van Cleef, Jessen and later Freeman, etc.).
Hartshorne reviewed Granö's ideas in some detail in
The nature of geography, but he could not bring
himself to accept Grano's view of the perceptual
environment as the basis for geographical study.
In Italy, too, Granö's views won a certain amount
of approval (Bertracchi, Renier).

Although Granö's ideas were closely associated
with philosophy and psychology they did not receive
much acclaim from leaders in those fields.   The
Professor of Philosophy at the University of
Helsinki, Eino Kaila, a man who trod the border area
between philosophy and psychology, discussed Grano's
theory of the perceptual environment, which he
regarded as notable, in his lectures in the 1930s
(oral communication from G.H. von Wright).   Granö's
influence can be seen in Kaila's posthumously
published work, *The perceptual and conceptual components of everyday experience* (1960).

In recent years Granö's contribution to geography has been reassessed by a number of authors (e.g.
Buttimer, Fischer, Hägerstrand, Kildema, Mead, Ohlson,
Ristkok, Vaart).   The writers have for the most part
assigned him neither to the German *Landschaft* school
nor to the school of artistic geography, which was
generally the case earlier.   Instead he has been

looked upon as a pioneer of perceptual geography.
The translation into English of Granö's 'Reine
Geographie' (in progress, 1978), which has been
instigated by Torsten Hagerstrand and Anne Buttimer,
is also a reflection of this new reassessment of
Granö's importance.   Consequently, there is now
a fresh wave of interest in Granö's environmental
geography among geographers, architects and environmental planners of his old countries, Finland and
Estonia.

Bibliography and Sources

1. REFERENCES ON J.G. GRANO
Merzbacher, G., 'Die Eiszeitfrage in der nordwest-
    lichen Mongolei', *Petermanns Geogr. Mitt.*, vol
    57,(1911), 18-19
Allix, A., 'Les glaciations quaternaires dans la
    Sibérie méridionale et la Mongolie occidentale',
    La Géogr., vol XXV,(1912), 431-7
Allix, A., 'Exploration du J.G. Granö dans la
    Mongolie occidentale', *La Géogr.*, vol XXVI, (1913)
Allix, A., 'L'Altai Russe', *Rev. Géogr. Alp.*, vol XI,
    (1923), 795-9
Tammekann, A., 'Prof. J.G. Grano maastiku-ja
    ümbruseteaduslised uurimised Valosaarel, Soomes'
    ('J.G. Granö's study of environment on the island
    Valosaari, Finland'), *Loodus*, vol 2, (1923), 641-9
Fickeler, Paul, 'Der Altai', *Petermanns Geogr. Mitt.
    Ergänzungsh.*, vol 187, (1925), 202 p.
Obruchev, W.A., 'Geologie von Sibirien', *Fortschr.
    Geol. u. Palaeont.*, vol 15 (1926), 1-572 (3,
    176, 391)
Kant, Edg., 'Tartu. Linn kui ümbrus ja organism'
    ('Tartu as environment and organism'), Tartu,
    1927, 192-433
Jaeger, Fritz, 'Granös Reine Geographie', *Petermanns
    Geogr. Mitt.*, vol 76 (1933), 67-8
Hassinger, H., 'Zur Methode der Landschaftsbeschrei-
    bung und Landschaftsgliederung' ('Description
    and regional division of the landscape'),
    Geogr. Z., vol 36, (1930), 293-6
Bertacchi, Cosimo,   'Reine Geographie', *Riv. Geogr.
    Ital.* vol XXXVII, (1930), 66-7
Van Cleef, Eugene, 'A philosophy of geographical
    regions', *Geogr. Rev.*, vol 22 (1932), 497-8
--   'Festschrift dedicated to J.G. Granö on his
    50th birthday', *Publ. Inst. Univ. Tartuensis
    Geogr.*, vol 20, (1932), 1-168
Kant, Edg.,   'Geographie, sociographie et l'ecologie
    humaine', *Publ. Sem. Univ. Tartuensis Oecon.-
    Geogr.*, vol 4, (1933), 1-61
Migliorini, E., 'Note geografiche sulle condizioni
    attuali dell'Estonia', *Boll. Soc. Geogr. Ital.*,
    vol 8-9, (1934), 559
Renier, Stefano, 'Un maestro e una scuola: J.G. Granö
    e l'Instituto geografico dell'Università di
    Abo', *Riv. Geogr. Ital.*, vol XLII, (1935), 24-8
Jessen, Jorgen, 'A Finnish geographer', *Baltic
    Countries*, vol 11, (1936), 1-6
Broek, J.O.M.,   'The concept landscape in human

geography', *C. R. Congr. Int. Géogr. Amsterdam,*
vol 2, (1938).103-9

Obruchev, W.A., 'J.G. Granö: Das Formengebäude des
nordöstlichen Altai, ('Geomorphology of NE.
Altai'), *Izv. Akad. Nauk SSSR Ser. Geol.,* vol 3,
(1947), 142-4

Chabot, Georges, 'J.G. Granö. Nécrologie', *Ann. Géogr.,*
vol LXV (1956), 313

Hildén, Kaarlo, 'Hommage à la mémoire de J.G. Granö',
Terra, vol 68 (1956), 2-13 (Includes a complete
bibliography)

Hildén, Kaarlo, 'J.G. Granö. Nécrologi', *Boll. Soc.
Geogr. Ital.,* vol IX, (1956), 601-2

Kant, Edg., 'Johannes Gabriel Granö. In memoriam',
Sven. Geogr. Årsbok, (1956), 163-172

Serebryannyj, L.E. and Gerasimov, I.P., 'J.G. Granö
1882-1956', *Izv. Akad. Nauk SSSR Ser. Geogr.,*
vol 6, (1956), 126-7

Hildén, Kaarlo, 'Johannes Gabriel Granö. Gedenkrede
(Obituary)', *Sitzungsber. Finn. Acad. Wissenschaften,* vol 1956, (1957), 57-69

Hang, E., 'J.G. Granö 1882-1956', *Eesti Geogr. Seltsi
Aastaraamat,* (1957), 257-8

Freemann, T.W., *A hundred years of geography,* London,
1961, 334 p.

Granö, Olavi, 'Piirteita J.G. Granöstä
tutkimusmatkailijana ja maantieteilijänä' ('J.G.
Granö, explorer and geographer'), *Terra,* vol 75,
(1963), 11-27 (Finnish with German summary 5 p.,
includes references on J.G. Granö).

Fischer, Eric, Campbell, Robert D. and Miller, Eldon S.
(ed), *A question of place,* Arlington, Virginia
1967, 446 p. (304-10)
Westermanns Lexikon der Geographie (Headwords:
Granö, Johannes Gabriel and Reine Geographie),
Braunschweig, 1969-70, 257, 987-8

Aalto, Pentti, *Oriental studies in Finland 1828-1918,*
Helsinki, 1971,, 174 p.

Kildema, K., 'J.G. Granö's heritage in the field of
landscape studies', *Eesti Loodus (Estonia Nature),*
vol May (1972), 295-8

Paffen, Karlheinz (ed), *Das Wesen der Landschaft*
('The nature of landscape'), Darmstadt, 1973,
514 p. (Collection of papers)

Van der Vaart, Jacob H.P., 'Grano's pure geography',
The Monadnock, vol June, (1975), 91-5

Winkler, Ernst (ed), *Probleme der allgemeine Geographie,* Darmstadt, 1975, 429 p. (Collection of
papers)

Ohlson, Birger, 'Sound field and sonic landscapes
in rural environment', *Fennia,* vol 148, (1976),
33-43

Schmithusen, Josef, *Allgemeine Geosynergetik,*
Berlin - New York, 1976, 349 p.

Mead, W.R., 'Recent developments in human geography
in Finland', *Progress in Human Geogr.,*
vol I, (1977), 361-75

Ristkok, A. 'On the theoretical basis of
estimating recreational characteristics of landscape with reference to J.G. Granö's heritage
on landscape science', *Eesti Geogr. Seltsi
Aastaraamat/Esogodnik Estonskogo Geogr. Obshch.
(Yearbook of the Estonian Geogr. Soc.),*
vol 1975/76, (1977), 162-84 (Estonian with
English and Russian summary 5 p.)

2. SELECTIVE AND THEMATIC BIBLIOGRAPHY OF WORKS BY J.G. GRANÖ

A complete bibliography appeared in *Terra,*
vol 68, (1956), 7-13.

a. Archaeology

1909 'Archäologische Beobachtungen von meinen Reisen
in den nördlichen Grenzgegenden Chinas in den
Jahren 1906 und 1907', *J. Soc. Finno-Ougrienne,*
vol XXVI, no 3, 1-54

1910 'Archäologische Beobachtungen von meiner Reise
in Südsibirien und der Nordwestmongolei im Jahre
1909', *J. Soc. Finno-Ougrienne,* vol XXVIII,
no 1, 1-67
'Über die geographische Verbreitung und die
Formen der Altertümer in der Nordwestmongolei',
J. Soc. Finno-Ougrienne, vol XXVIII, no 2, 1-55

1958 'Materialien zu den alttürkischen Inschriften der
Mongolei' (with G.J. Ramstedt and Pentti Aalto),
J. Soc. Finno-Ougrienne, vol 60, no 7, 1-91

b. Glacial geology and geomorphology

1910 'Beiträge zur Kenntnis der Eiszeit in der nordwestlichen Mongolei und einigen ihrer südsibi-
rischen Grenzgebirge', *Fennia,* vol 28, no 5,
1-230 (thesis)

1915 'O lednikovom periode v Russkom Altaje' ('Glacial
period in the Russian Altai'), *Izv. Zapad.
Sibirsk. Russ. Geogr. Obshch.,* vol 2, 1-58

1916 'O znachenyi lednikovago perioda dla morphologyi
Severovostochnovo Altaja', ('Influence of
Pleistocene on morphology of the NE. Altai'),
Izv. Zapad. Sibirsk. Russ. Geogr. Obshch., vol 3,
1-22

1917 'Les formes du relief dans l'Altai russe et leur
genèse. Étude morphologique', *Fennia,* vol 40,
no 2, 1-125

1945 'Das Formengebäude des nordöstlichen Altai'
('Geomorphology of NE. Altai'), *Publ. Inst. Univ.
Turkuensis. Geogr.,* vol 20, 1-362

1950 'Les terrasses dans les vallées de l'Altai
nord-est, témoins de l'évolution morphologique
de ces montagnes', *C.R. XVI Congr. Int. Géogr.
Lisbonne 1949,* 482-6

1952 'Land Forms and relief', *Suomi. A general handbook on the geography of Finland, Fennia,*
vol 72, 74-99

c. Philosophy and methodology of geography and regional geography

1909 'Piirteitä eläinmaantieteen historiasta, etenkin
aluejakokysymystä silmällä pitäem' ('History of
zoogeography with special reference to regional
division'), *Suome Maantieteellisen Yhdistyksen
Julkaisuja (Geogr. Soc. Publ.),* vol 8, 1-93
(Finnish with German summary 2p.)

1910 'Maantieteellisen maakuntajaon perusteista' ('The
bases of regional division'), *Maantieteellisen
Yhdistyksen Aikakauskirja (Geogr. Soc. Publ.),*
vol 22, 309-14 (Finnish with German summary 2p.)

1919-21 *Altai, vaellusvuosina nähtyä ja elettyä, I-II
(Altai, seen and experienced during years of travel),*
Porvoo, 708 p. (Finnish and Swedish)

1920 'Maantieteestä, sen asemasta yliopistossamme ja
sitä meillä edustavista seuroista' ('Geography
and its status in Helsinki University and the
Geographical Societies in Finland'), *Terra*,
vol 32, 13-50 (Finnish and Estonian with German
summary in *Petermanns Geogr. Mitt.*, vol 66, 229)

1922 'Eesti maastikulised üksused' ('Landscape units
in Estonia'), *Loodus*, vol 1, 105-23, 193-214,
257-81 (Estonian with German summary 23 p.)

1923 'Maastik ja ümbrus' ('Landscape and environment'), *Loodus*, vol 2, 321-7 (Lecture notes by
Edg. Kant, Estonian)

1927 'Die Forschungsgegenstände der Geographie' ('Geo-
graphy's object of study'), *Acta Geogr.*, vol 1,
no 2, 1-15 and *Sven. Geogr. Årsbok*, 34-46
(German and Swedish, lecture held in Lund,
Copenhagen, Stockholm, Hamburg and Greifswald)
'Suomalainen maisema' ('Finnish landscape'),
Terra, vol 39, 73-85 (Inaugural lecture, Turku,
with German summary 2 p.)

1928 'Geographical regions', *Atlas of Finland 1925*
and *Fennia*, vol 48, 116-33

1929 'Reine Geographie', *Acta Geogr.*, vol 2, no 2,
1-202 (German and Finnish)

1930 'Maisematieteellinen aluejako' ('Regional
division of landscape'), *Terra*, vol 42, 1-26
(Finnish with German summary 4 p.)

1931 'Die geographische Gebiete Finnlands', *Fennia*,
vol 52, no 3, 1-182 (German and Finnish)
'Altain kasvillisuus' ('Vegetation of Altai'),
Terra, vol 43, 1-20 (Finnish with German
summary 4 p.)

1935 'Geographische Ganzheiten', *Petermanns Geogr.
Mitt.*, vol 81, 295-302 (New edition in *Das
Wesen der Landschaft*, Darmstadt, 1973, 3-19)

1938 'Itinerarien und Landschaftsprofile J.G. Granös
aus Uranchai (Tunnu-Tuwa) und Nordmongolei'
(with A.K. Merisuo and Pentti Eskola), *Acta
Geogr.*, vol 6, no 1, 1-54

1941 'Mongolische Landschaften und Ortlichkeiten'
('Mongolian landscapes and regions'), *Acta
Geogr.*, vol 7, no 2, 1-291

1947 'Aluemaantieteellisestä tutkimuksesta ja sen
tehtävistä maassamme' ('Regional geographical
studies and their applications'), *Terra*, vol
59, 25-31 (Finnish with German summary 1 p.)

1952 'Régions géographiques et une méthode pour les
délimiter', *C.R. XVI Congr. Int. Géogr.
Lisbonne 1949*, 322-31
'Geographic regions', *Suomi. A general handbook on the geography of Finland*, *Fennia*,
vol 72, 408-38

1955 'Die geographische Provinzen Finnlands', *Geogr.
Rundschau*, vol 7, 81-94

d. Settlement and urban geography

1905 'Siberian suomalaiset siirtolat' ('Finnish
colonies in Siberia'), *Fennia*, vol 22, no 4,
1-86 (Finnish with German summary 22 p.)

1923 'Hembygdsforskning i Estland' ('Local research
in Estonia'), *Terra*, vol 35, 131-6 (Swedish
and Finnish).

1928 'Vegetation and distribution of population',
Atlas of Finland 1925 and *Fennia* 48, 81-91

1937 'Gehöfte und Siedlungen in Finnland'
('Dwellings and settlements in Finland'),
Fennia, vol 63, no 6, 1-66

1952 'Settlement of the country', *Suomi. A
general handbook on the geography of
Finland*, *Fennia*, vol 72, 340-80

*Olavi Granö is Professor of Geography at Tuku
University, Finland.*

CHRONOLOGICAL TABLE: JOHANNES GABRIEL GRANÖ

DATES	LIFE AND CAREER	ACITIVITIES, TRAVEL, FIELDWORK	PUBLICATIONS	CONTEMPORARY EVENTS AND PUBLICATIONS
1882	Born 13 March at Lapua, Ostrobothnia, W. Finland			*Anthropogeographie* (F. Ratzel)
1883				Finno-Ugric Soc. founded
1885	Family moved to Omsk, W. Siberia			*Aufgaben und Methoden der Geographie* (F. von Richthofen)
1886				*Landschaftskunde* (A. Oppel); *Führer der Forschungsreisende* (F. von Richthofen)
1888				*Geografins uppgift* (R. Hult); Geogr. Soc. of Finland founded
1889				Department of Geography founded at Helsinki Univ.
1890				*Geography of Finland* (K.E.F. Ignatius); first docent in geography (R. Hult)
1891	Family returned to Finland			*Lage der Geographie* (E. Löffler)
1892	Attended school at Tornio, N. Finland			Geography established as examination subject
1894				O. Donner President of Finno-Ugric Soc.
1895				*Finlands naturliga landskap* (R. Hult)
1896	Attended secondary school at Oulu, N. Finland			
1897				*Zeitströmungen in der Geographie* (E. Wisotzki)
1899				*Atlas of Finland*, 1st ed
1900	Entered Helsinki Univ.			*Lehrbuch der Geographie* (H. Wagner)
1901	Began studies in geography			First professor of geography (J.E. Rosberg); *Altai* (W. Saposnikov)
1902	Family moved to Omsk once more	Summer travels in W. and S. Siberia		*Antlitz der Erde* (E. Suess)
1903	Assistant in Geography at Helsinki Univ.			
1904				*Tian-Shan* (W.M. Davis)
1905		Journey to Altai	'Finnish colonies in Siberia'	*Wesen und Methoden der Geographie* (A. Hettner); *The major natural regions* (A.J. Herbertson)

DATES	LIFE AND CAREER	ACTIVITIES, TRAVEL, FIELDWORK	PUBLICATIONS	CONTEMPORARY EVENTS AND PUBLICATIONS
1906	M.A. degree, Helsinki Univ.; secondary school teacher	Exploration in NW.Mongolia, Uriankhai and Dzungaria; working member, Geogr. Soc. of Finland		*Ziele der Geographie* (O. Schlüter)
1907		Exploration in Altai, NW. Mongolia and Uriankhai		
1908	Lecturer in Geography at Helsinki Univ. (to 1919)			*Landschaften Afrikas* (S. Passarge); *Einteilung der Erdoberfläche* (A. Hettner)
1909		Exploration in S.Siberia, Mongolia and Trans-Baikal	'Archäologische Beobachtungen 1906-7'; 'History of zoogeography'	*Alpen im Eiszeitalter* (Penck and Bruckner)
1910			'Eiszeit in der NW. Mongolei', thesis; 'Archäologische Beobachtungen 1909'	*Atlas of Finland*, 2nd ed
1911	Ph.D. degree Helsinki; docent in geography	Journey to E.Siberia and Japan		*Geopsychische Erscheinungen* (W. Hellpach); *Naturschilderung* (F. Ratzel)
1912		Assoc. member, Finnish Academy of Sciences	'NW.Mongolei'	*Geografins mal och medel* (J.J. Sederholm); *Physiologische Morphologie* (S. Passarge); *Geographie* (E. Banse); *Erklärende Beschreibung der Landformen* (W.M. Davis)
1913	Married Hilma Ekholm, M.A., moved to Omsk (grant for 1913-15)	Journey to Crimea, Caucasus, Tien-Shan (with Rosberg) and Altai		*Caractères distinctifs de la geographie* (P. Vidal de la Blache)
1914		Exploration in Altai; member, Russian Geographical Society		*Altai* (W. Obruchev)
1915		Exploration in Altai	'Glacial period in the Russian Altai'	*Subject and aims of geography* (L.S. Berg)
1916		Exploration in W.Altai	'Influence of Pleistocene on morphology of the NE.Altai'	
1917	Geogr. editor of Finnish Encyclopaedia and regional monographs of Finland		'Formes du relief dans l'Altai'	Finland and Estonia independent; *Geologische Bau und Landschaftsbild* (K. Sapper); *Wesen Mitteleuropas* (H. Hassinger)
1918				War in Finland
1919	Professor of Geography, Tartu Univ., Estonia		'Altai' I (Travelogue)	*Einheit der Geographie* (A. Hettner); Estonian Univ. founded anew at Tartu

DATES	LIFE AND CAREER	ACTIVITIES, TRAVEL, FIELDWORK	PUBLICATIONS	CONTEMPORARY EVENTS AND PUBLICATIONS
1920		Founder of Estonian Local and Regional Research Soc.	'Geography and its status'	Turku Univ. founded; *Expressionismus und Geographie* (E. Banse)
1921		Member, American Geographical Society	'Altai' II (Travelogue)	*Landschaftskunde* (S. Passarge)
1922	Editor, *Atlas of Finland*	Member, Finnish Scientific Society	'Landscape units in Estonia'	
1923	Professor of Geography, Helsinki Univ.	Corresponding member, Geographical Society of Berlin	'Landscape and environment'	
1924		Member, Finnish Academy of Sciences		*Morphologische Analyse* (W. Penck); *Harmonische Landschaftsbild* (R. Gradmann)
1925				*Geist der Wissenschaft* (F. Neef); *Géographie humaine* (J. Brunhes)
1926	Professor of Geography, Turku Univ.	Lecture tour of Scandinavian universities; honorary member, R. Danish Geographical Society; editor, *Z. Geomorphologie*		*Geologie von Sibirien* (W. Obruchev) *Geographie* (H. Lautensach)
1927		Corresponding member, Swedish Geographical Society	'Geography's object of study'; 'Finnish landscape'	*Geographie* (A. Hettner)
1928			*Atlas of Finland*	*Landschaft und Seele* (E. Banse)
1929		Paper at 18th Scandinavian Naturalist Congress in Copenhagen	'Reine Geographie'	
1930		Lecture tour to Hamburg and Greifswald	'Regional division of landscape'	
1931			'Geographische Gebiete Finnlands'; 'Vegetation of Altai'	*Kritik der wissenschaftlichen Grundbegriffe* (H. Cornelius)
1932	Rector, Turku Univ. (to 1934)	Honorary member, Estonian Natural Society		
1933		Doctor, *honoris causa*, Tartu, Estonia; honorary member, Pomeranian Geographical Society		
1934	Resigned from post of Rector due to ill health	Honorary member, Estonian Society of Letters	Paper at 14th Int. Geogr. Congress, Warsaw	*Wahrnehmung und Gegenstandswelt* (E. Brunswik); *Grundprobleme der Philosophie* (H. Rickert)
1935			'Geographische Ganzheiten'	
1936		Honorary member, Geographical Society of Leipzig		

DATES	LIFE AND CAREER	ACTIVITIES, TRAVEL, FIELDWORK	PUBLICATIONS	CONTEMPORARY EVENTS AND PUBLICATIONS
1937			'Dwellings and settlements in Finland'	
1938		President, Geographical Society of Finland (50th Anniversary); honorary member, Geogr. Soc. of Latvia; Geogr. Soc. of Munich; Geogr. Soc. of Stettin	'Itinerarien und Landschaftprofile aus Uranchai und N.Mongolei'	15th Int. Geogr. Congress, Amsterdam
1939				*Nature of Geography* (R. Hartshorne); War breaks out in Finland
1940		Honorary member, Finnish Antiquarian Society		
1941			'Mongolische Land-schaften'	
1942		Honorary member, Geographical Society of Finland		
1945	Professor of Geography, Helsinki Univ.; Chancellor, Turku Univ. (to 1955)		'Formengebäude des nordöstlichen Altai'	End of the war in Finland
1946		Honorary member, Geographical Society of South Sweden		
1947		Fennia Gold Medal of Geogra-phical Society of Finland	'Regional geographi-cal studies and their applications'	
1948		Chairman, Finnish Academy of Sciences		
1949		Papers at 16th Int. Geogr. Congress, Lisbon and excursion to Madeira		
1950	Retirement from post of Professor	Honorary member, Geographical Society of Lisbon	'Terrasses de l'Altai'	
1951		Editor, *Handbook on the geo-graphy of Finland*	Chapters in *Handbook on the geography of Finland*	
1952		Chairman, Editorial Board, Atlas of Finland; honorary member, Finnish Cultural Foundation	'Régions géographiques'	
1953		President, 3rd Congress of Science, Helsinki		
1954		Honorary member, Geographical Society of Berlin		
1955	Resigned from post of Chancellor	Doctor, *honoris causa*, Turku	'Geographische Provinzen Finnlands'	
1956	Died 23 February in Helsinki			

Andrew John Herbertson

1865–1915

L. J. JAY

1. EDUCATION, LIFE AND WORK

Born at Galashiels in Selkirkshire, Andrew John
Herbertson was the eldest of four children.   Their
father, a staunch Presbyterian, a Justice of the Peace
and an active Liberal in politics, was a prosperous
building contractor who owned several quarries in the
neighbourhood; his wife was the daughter of a hill
farmer from the head of Peel Glen.   After attending
the Academy in Galashiels Andrew moved to Edinburgh to
spend two years at the Institution, now known as
Melville College, where his course emphasized science,
mathematics and modern languages.   Before proceeding
to university he spent some time acquiring practical
experience with a firm of surveyors in the city.

He was almost 21 before he registered as a
student at the University of Edinburgh, and although
he studied there for four years, with an interval
spent at the University of Freiburg-im-Breisgau, he
never graduated from Edinburgh. He always registered
as an Arts student, yet he chose to attend science
classes, although curiously enough he ignored
chemistry which was required for the B.Sc. degree
course. .Nevertheless, despite his unorthodox
approach, he impressed his tutors sufficiently to win
two valuable scholarships in physics. He attended
mathematics classes conducted by Professor Chrystal
and physics lectures given by P.G. Tait and studied
geology under James Geikie.   He also became a fervent
admirer and friend of Patrick Geddes, whose versatile
interests ranged over biology, sociology and town-
planning.   Herbertson acted as assistant lecturer at
the popular Summer Schools which Geddes organized for
several years in Edinburgh, and it was at one of these
that he met Dorothy Richardson, a classics mistress
at Cheltenham Ladies' College, who had become
interested in sociology.   They were married in the
autumn of 1893 and spent the ensuing winter in France,
where Andrew worked at the meteorological station in
Montpellier and also with the botanist Flahault in
Paris.   On their return to Britain in 1894 Herbertson
took up a post as lecturer at Owen's College, University of Manchester, but within two years he
returned to Edinburgh to the Heriot-Watt College.
Both of these appointments were held before he had
acquired a degree.

The closing years of the century were packed with
activity for Herbertson.   In addition to preparing
reports for the Scottish Meteorological Society on
his hygrometric researches at the Ben Nevis Observatory
he delivered the first of a long series of pronouncements on geography in education at the Sixth International
Geographical Congress, held in London in 1895.   He
successfully presented his Ph.D. thesis at Freiburg,
edited and contributed substantially to an *Atlas of
meteorology* in collaboration with J.G. Bartholomew
under the general supervision of Alexander Buchan, and
acquired editorial experience with two geographical
journals, the *Scottish Geographical Magazine* and the
American *Journal of School Geography*.   He was elected
to the committee of the Geographical Association and
wrote, with his wife, one of the most successful of
their numerous textbooks for schools, *Man and his
work*.   To crown all these accomplishments he was in
1899 persuaded by H.J. Mackinder to become Lecturer in
Regional Geography in the newly created School of
Geography at the University of Oxford, where he

remained for the rest of his life. In 1905 he
succeeded Mackinder as Reader in Geography and Head
of the School.

His output of contributions to learned journals
as an Oxford don was less impressive in bulk than
the profusion of textbooks, articles, reviews and
editorial items relating to education and the
teaching of geography in school which he produced.
The Geographical Association was only seven years
old when Herbertson became its Honorary Secretary,
and he was joint or sole Editor of its journal,
The Geographical Teacher, from its inception in 1901
until his death. For fifteen years the Association
grew in stature and strength, largely in consequence
of his tireless efforts and shrewd leadership.

Of medium height and sturdy frame, with blue
eyes and a fresh complexion, Herbertson lacked the
compelling magnetism of Mackinder as a lecturer, but
he was a patient adviser in tutorials and a
stimulating director of practical work in the field.
In addition to the normal course for the Diploma
in Geography, the Oxford School became famous for
its biennial Summer Schools which attracted large
numbers of teachers and several distinguished
visiting lecturers. Herbertson, aided by his wife,
rose magnificently to these occasions, for his
academic ability went along with a measure of
practical business efficiency.

Bouts of ill health became more frequent following the outbreak of war and he died at his country
home in the Chilterns on the last day of July 1915.

2. SCIENTIFIC IDEAS AND GEOGRAPHICAL THOUGHT

a. The scientific contibution

Herbertson used the Neil Arnott Scholarship in
Experimental Physics, which he was awarded in 1887,
to assist Professor Tait in the construction of
impact apparatus. Five years later, when he received a valuable Physical Science Scholarship, he
acted on advice from Tait, whose interest in meteorology he had absorbed, and used the grant to conduct
highlevel hygrometric observations at the Ben Nevis
Observatory. Although the atmosphere at the summit
of this mountain is generally moistureladen, there
are occasional periods of fine dry weather when the
air is much drier that it is at lower altitudes.
Herbertson sought to determine from readings of the
wet and dry bulb thermometers the pressure of the
water vapour in the air and the relative humidity
during these rare dry spells at the summit. He
worked at the high-level Observatory on Ben Nevis for
five months during the winter of 1892-3 and was
fortunate to experience one of the exceptional spells
of dry atmosphere over Christmas and the New Year.
He used the laborious but precise gravimetrical
method to obtain data for constructing hygrometrical
tables suited to dry conditions at very low tempera-
tures. His investigations were thorough and his
findings, subsequently published in various issues of
the *Journal of the Scottish Meteorological Society*
and the *Transactions and Proceedings of the Royal
Society of Edinburgh*, were models of lucidity.

During the following year he worked at the low-
level observatory at Fort William, and also pursued
meteorological investigations at Montpellier
and in Dr Friedel's laboratory in Paris. Working
in collaboration with J.G. Bartholomew he was partly
responsible for an impressive *Atlas of meteorology*
and around this time he completed a Ph.D. thesis on
the subject of 'Monthly rainfall over the land surface of the globe'. This was accepted *multa cum
laude* by the University of Freiburg in 1898 and was
published three years later as a special monograph of
the Royal Geographical Society under the title of
'The distribution of rainfall over the land',
with coloured maps showing, for the first time, the
mean rainfall for each month of the year over the
globe.

Surprisingly, this proved to be the last
publication of Herbertson's fieldwork in meteorology
and hydrology, except for one belated report on his
work at the Ben Nevis Observatory, which was not
printed until 1905 in the *Transactions of the Royal
Society of Edinburgh*. However, his investigations
into the global distribution of rainfall and
temperature, coupled with the study of plant-
associations which had been stimulated by Geddes in
Scotland and Flahault in France, led Herbertson to
formulate the concept of major natural regions which
he presented at a meeting of the Research Committee
of the Royal Geographical Society in February 1904.
Despite the clarity of presentation his paper
was sharply criticized by H.R. Mill while other
participants in the discussion such as Freshfield,
Ravenstein and Mackinder merely echoed the querulous
tone of Mill. The printed version appeared in vol
25 of the *Geographical Journal* (1905), 300-12, under
the title 'The major natural regions: an essay in
systematic geography' and it became the most widely
read of Herbertson's contributions to learned
journals. In making a rational subdivision of the
lands of the world into a dozen or so type-areas
Herbertson favoured climate as the fundamental
factor, although from the start he conceded that the
distribution of vegetation and even human beings
might also be worthy of consideration. For the
structural subdivisions he drew on the monumental
work of Suess, *Das Antlitz der Erde*, and on seasonal
rainfall he acknowledged the help given by the
maps in Supan's *Grundzüge der physischen Erdkunde*
(1896). His seasonal variations of temperature
were based on the mean monthly sea-level isotherms
of 0°, 10° and 20°C for the coldest month and 10°
and 20°C for the warmest month, a scheme which
simplified criteria used by Köppen in an article
published in the *Geographische Zeitschrift* in 1900.
Herbertson defined a major natural region as one
which possessed 'a certain unity of configuration,
climate and vegetation' and he recognized that the
boundaries of the regions were rarely defined as
sharply as were 'political' frontiers. The map
to accompany his original paper showed fifteen major
regional types, but in the *Senior geography* (1907)
he combined the Monsoon type and Sudan type to form
a subdivision of the Hot Lands characterized by
summer rain, thereby reducing his generic regions
to fourteen. He also had second thoughts about the
criteria for temperature belts, which originally had
been based on sea-level isotherms. In a paper
published in 1912 entitled 'The thermal region of the
globe' (*Geogr. J.*, vol 40, 518-32) he stated that 'in

all geographical work we find that the isotherms reduced to sea-level are insufficient', and his substitution of actual isotherms placed greater emphasis on the influence of relief in determining natural regions.

Another refinement to his original scheme which he subsequently contemplated was to give more attention to biological factors. 'If the geographical region is a macro-organism then men are its nerve cells...for the full understanding of the organism, or of the macro-organism, the nervous system, or the human society, cannot be separated from it' ('The higher units', *Scientia*, vol 14, 1913, 212). The map on p. 208 of this paper reveals the modifications which Herbertson had made to his original classification of major regions. He was still preoccupied with the human factor in the geographical complex when he died, for in an incomplete fragment, which was printed posthumously in the *Geographical Teacher* (vol 7, 1915, 147-53) he referred to a *genius loci*, or spirit of place, which alters with man's relation to the region ('Regional environment, heredity and consciousness'). It was perhaps unfortunate that in his school textbooks he omitted to provide details of the criteria by which his regions were revised, or to reveal the increasing importance which he attached to human factors, for this led many teachers to regard 'natural' as synonymous with 'physical' in the definition of his regions, whereas Herbertson intended the word as the antonym of 'artificial' or 'arbitrary'.

b. *Ideas held on geography and other sciences*

In the days before geography was considered to be a fit subject for study in the Universities of Oxford and Cambridge there was no uniform course of education for those men and women who subsequently helped to re-fashion the aims and scope of the subject so that it acquired status in the academic world. Thus Keltie approached geography through journalism, H.J. Mackinder entered from history, H.R. Mill from chemistry and G.G. Chisholm from statistics. Herbertson was exceptional in the breadth of his youthful interests which shaped his outlook as a geographer, and in his unorthodox selection of studies at Edinburgh. He attended classes in mathematics, geology, agriculture and astronomy before concentrating his attention on physics and meteorology. Meanwhile at Freiburg he studied geography, meteorology and oceanography under Professor Neumann. (Keltie observed in his Report of 1885 that Germany possessed twelve full professors of geography at a time when Britain had none.) From Patrick Geddes, Herbertson acquired a compelling interest in biology and sociology and was introduced by him to the works of Frenchmen eminent in these fields - the botanist Flahault, the sociologists Demolins, de Rousiers and Le Play, and the geographer Élisée Reclus. Visits to France and Germany strengthened his understanding of the written and spoken languages of these countries, and enabled him to read the works of continental geographers in their original tongue. (The first book which he published was a translation into English of *La vie américaine* by Paul de Rousiers, in 1892.)

Another powerful influence in Edinburgh in the 1880s was John Murray, who had been one of the naturalists on the *Challenger* expedition. Through him Herbertson acquired practical knowledge of hydrology in varied forms - making physical observations of the seas bordering Scotland, sounding the depths of English lakes, and collecting hygrometric data from the summit of Ben Nevis. This experience was subsequently put to good use in 1906-7 when he became a member of the Royal Commission on Canals and Inland Waterways and the Committee on Water Supply.

Patrick Geddes had picked up ideas about the regional method of teaching from T.H. Huxley, whose *Physiography* (1877) he considered to be 'surely one of the best elementary textbooks ever written', and the Outlook Tower on Castlehill, Edinburgh, was developed to become a laboratory for the demonstration of regional survey in action; it was furnished with two large globes, one of relief and another of vegetation, which were constructed by Herbertson and Reclus. Mackinder stressed the value of the regional approach in his famous lecture in 1887 'On the scope and methods of geography' (*Proc. R. Geogr. Soc.*, vol 9, 141-60), while H.R. Mill edited *The international geography* in 1899 and wrote the section in it on Great Britain, on a regional basis. In 1903 Vidal de la Blache produced his *Tableau de la géographie de la France*, which dealt with the small regional units or *pays* of the Republic. Thus when Herbertson moved to Oxford in 1899 as Lecturer in Regional Geography he was aware of this approach, and approved of it. One element of regional study was the detailed examination and description of a small area familiar to the student, and for the Oxford Diploma this exercise involved surveying the area covered by a sheet of the one-inch Ordnance Survey series. As an illustration of the technique, Herbertson himself published an account of the Oxford Sheet, in the *Geographical Teacher* vol 1, 1901-2, 150-66; he acknowledged the inspiration given by H.R. Mill's monograph on south-west Sussex. However, this device of studying the interactions of geographical phenomena in regions of sizes varying from a small valley to a nation-state should not be confused with another concept, which Herbertson had been pondering for several years. He disliked two features of the geography taught in school during the nineteenth century: the separation of subject-matter into distinct compartments labelled 'physical' and 'political', and the strain placed upon a child's memory by using a multitude of administrative divisions as units of study. Herbertson sought to reduce the mass of information which formed the content of geography by devising a limited number of type-areas, and this search led him to formulate the concept of major natural regions. He was employing this idea in lectures to his students at Oxford some years before, as he indicated in his paper 'Geography in the university' (*Scott. Geogr. Mag.*, vol 18, 1902, 128).

c. *The world view*

By ignoring the political boundaries between countries when he delimited the extent of his major natural regions, Herbertson was criticized by some geographers for failing to take account of the effects of national policy on geographical development. L.W. Lyde favoured political units for study, and modern trends seem to support this preference. *Europe, a regional geography* was written by N. MacMunn and G. Coster in 1922 from notes left by Herbertson, but in spite of certain merits the book followed a pattern which was not widely copied. On the other hand, there was more support for Herbertson's belief that the study of geography should be used to assist the understanding of economic and social problems and promote a sympathetic attitude towards other countries than one's own. To *The practice of instruction*, edited by J.W. Adamson (1907), Herbertson contributed a lengthy section on 'geography', which concluded with the hope that it would yield a true appreciation of the conditions and needs of the home region, and of its relation to the world beyond the home district. Properly used, he wrote, geography should foster first a local patriotism, then an affection for country and Empire, and ultimately assume the loyalty of a citizen of the world. Sentiments of this nature appeared in the Editorial Notes which he wrote for the *Geographical Teacher* as the threat of war became a reality. After his death J. Cossar, commenting on Herbertson's contributions to the study of geography, declared that Herbertson expounded the ideal of synthesis, of sympathy and of mutual cooperation which he hoped would dominate the activities of man and mould his development in the twentieth century, just as the ideals of analysis and individual struggle had directed human progress during the latter half of the previous century (*Scott. Geogr. Mag.*, vol 31, 1915, 486-90). In 1922 an American observer regarded Britain as the country which, to a greater degree than elsewhere in Europe, was making geography serve as a medium for the more sympathetic understanding of other peoples, and he considered that Herbertson had been one of the leading exponents of this view (W.L.G. Joerg, 'Recent geographical work in Europe', *Geogr. Rev.* vol 12, 1922, 432).

3. INFLUENCE AND SPREAD OF IDEAS

It was remarkable how rapidly the scientific work in meteorology and climatology, which Herbertson conducted towards the end of the nineteenth century, was forgotten in the twentieth, so much so that when Herbertson died Mackinder could boldly assert 'nine out of ten who have known Herbertson will remember him...for his great efforts in connection with geographical education rather than geographical science' (*Geogr. Teach.*, vol 8, 1915, 144). This reflects the change in emphasis which Herbertson gave to his work after moving to Oxford.

Mackinder adopted the regional method as a device to merge the physical and political aspects of geography which had sundered during the nineteenth century, and this method figured prominently in the syllabus for the Oxford Diploma in Geography after 1899. As J.N.L. Baker has demonstrated, in *The history of geography* (p. 126), Mackinder was not the first man to preach the unity of geography, nor

was he the pioneer of the regional method, but his brilliance as a lecturer gave this concept the publicity which hitherto it had lacked. The alternative approach to the regional treatment of geographical data was called 'systematic', and W.M. Davis outlined these contrasted methods of study in 'A scheme of geography' (*Geogr. J.*, vol 22, 1903, 413-23). It was because of this special significance which geographers attached to the word 'systematic' that Mackinder, with some justification, objected to Herbertson's use of this word in the full title of his 1904 paper on the grounds that it was confusing.

When Herbertson began lecturing at Oxford he endorsed and employed the regional method, but his concept of major natural regions was more than a device for merging physical and human ('political') geography; in effect it was a simplified framework on which global studies could more easily be conducted. Although his concept was frequently addressed to adult audiences or readers, one suspects that Herbertson was often thinking in terms of geography at a less mature level. Thus in the article 'The higher units', written for scientists in 1913, he refers more than once to 'the pupil' and 'the child' on p. 211, which suggests that he was visualizing the teaching of geography in school. This may well be so, for during the last 16 years of his life his talents were directed to the advancement of geography teaching with a startling intensity. The Clarendon Press sold nearly one and a half million copies of the 'Oxford geographies', a series of school books written by Herbertson and his wife incorporating his ideas on regions and regional method, while *Man and his work*, first published by Black in 1899, was revised several times, the latest edition appearing as recently as 1963. Apart from his school books Herbertson influenced many teachers by the numerous articles, reviews and editorial items which he wrote for the *Geographical Teacher* and *School World* and other educational publications; these are classified in a list of his published works in *Geography* vol 50, 1965, 364-70. Moreover, he stimulated others to write successful textbooks: J.B. Reynolds, Marion Newbigin, J.F. Unstead, E.G.R. Taylor, L. Brooks, C.B. Thurston, and L. Dudley Stamp are names familiar to the generation which attended school between the two World Wars, and all of these authors have acknowledged their debt to Herbertson. He joined the Oxford School of Geography at its inception in 1899, and was its Head for ten of its first 16 years. J.N.L. Baker considered that Herbertson and Dickson contributed more than Mackinder to the Oxford school, which before 1920 had furnished the whole of the teaching staff to six departments of geography in universities or university colleges, and one or more members to the staffs of six more departments (*The history of geography*, 1963, 82 and 128). The University of Oxford recognized his worth by conferring on him the personal title of Professor in 1910. The only other professor of geography in Britain was L.W. Lyde who had been appointed to a chair at University College, London, in 1903. E.W. Gilbert proposed the formation of a Herbertson Society which was established at Oxford in 1923; essentially an undergraduate body but with a section comprising life-members, its activities include lectures, discussions and an annual excursion. A Herbertson Prize, initiated in 1921, is awarded each year on the results of the examinations in the Oxford School of Geography.

The Geographical Association increased its membership ten-fold during the years when Herbertson was its Honorary Secretary and Editor, and his guidance through a formative period of its existence was of incalculable benefit. From his election to the committee of the Association (designated 'Council' after 1913) until his death 17 years later, Herbertson missed only five of its fifty meetings, and his influence permeated every aspect of the Association's activities and publications. The esteem in which he was held led to the foundation of the Herbertson Memorial Lecture in 1917, to be given 'by some serious contributor to geographical knowledge or interpretation'; the fifteenth of these Lectures was delivered in 1973. The centenary of Herbertson's birth in 1965 was commemorated in a Lecture given by Professor E.W. Gilbert, who delivered a masterly appreciation of Herbertson's life and work, and this was printed in a special Herbertson Centenary issue of *Geography*, which also contained tributes from H.J. Fleure and J.F. Unstead, a bibliography of Herbertson's published works, and a list of the Memorial lectures (*Geography*, vol 50, 1965, 313-72).

In academic circles the name of Herbertson is associated with the concept of major natural regions, of which the merits have been debated ever since it was first put forward in 1904. Anyone attempting to assess its significance must examine it relative to the state of geography as it was at the beginning of the twentieth century, not as it exists today, for the subject has evolved over the years, and the regional concept is no longer adequate to meet all demands. It must be remembered that Herbertson deliberately intended to devise a simple framework for global studies, and that he was constantly modifying his original scheme.

Among university geographers the effect was more in terms of inspiration rather than in wholehearted adoption of the concept. Towards the end of his life Herbertson was giving more attention to the types and orders of natural regions, and J.F. Unstead developed a regional hierarchy by proceeding from the smallest unit-areas, which he called 'stows', to progressively larger areas formed by synthesis until he reached major regions. ('A synthetic method of determining geographical regions', *Geogr. J.*, vol 48, 1916, 230-49). P.M. Roxby drew attention to the importance of space-relations ('The theory of natural regions', *Geogr. Teach.*, vol 13, 1926, 376-82). H.J. Fleure, following another line of thought, subdivided Europe into regions according to the degree of human effort expended in them; he thus defined zones of Increment, Effort, Difficulty and Privation ('Regions in human geography, with special reference to Europe', *Geogr. Teach.*, vol 9, 1918, 31-45).

Herbertson was probably more familiar with the writings of German geographers than were most of the British contemporaries, and his reputation stood high in many continental universities. Professor Friedrich of Leipzig analyzed the economic significance of Herbertson's regions in 1911, in his

Geographie des Welthandels und Weltverkehr. Ten years later S. Passarge paid a handsome tribute to Herbertson's work in the Introduction to his comparative regional geography, *Vergleichende Landschaftskunde* (1921). E.W. Gilbert, in his Centenary Lecture delivered in 1965, drew attention to an article by Professor Willi Czajka of the University of Göttingen, in which Herbertson's concept is applauded and related to the ideas of German, American and British geographers. ('Die geographische Zonenlehre: 50 Jahre nach A.J. Herbertson's "The major natural regions"', *Geogr. Taschenb.*, Wiesbaden, 1957, 410-29).

Andrew Herbertson was a reticent person who did not seek the plaudits of the academic world. As an undergraduate he scorned the orthodox examination procedure; as a lecturer he did not deem it necessary to scale an African peak, as Mackinder did in 1899, merely to win acceptance among his fellow geographers. His firm resolve to concentrate his talents on improving the status of geography in schools may have reduced his prestige in the universities, and it undoubtedly diminished his opportunities for conducting the advanced scientific research of which he had proved himself capable. Nevertheless he exerted a powerful influence on the theory and practice of geography in Britain during the first half of the present century.

Bibliography and Sources

1. REFERENCES AND SPECIAL STUDIES ON A.J. HERBERTSON
a. Obituaries
Beckit, H.O., *Geogr. J.*, vol 46, (1915), 319-20
Cossar, J., *Scott. Geogr. Mag.*, vol 31, (1915), 486-90
Watt, A., *J. Scott. Meteor. Soc.*, vol 17, (1915), 34-6
Mackinder, H.J., MacMunn, N.E., Elton, E.F., *Geogr. Teach.*, vol 8, (1916), 143-6

b. Articles in Geography *vol 50, (1965), Herbertson Centenary Number*
Fleure, H.J., 'Recollections of A.J. Herbertson', 348-9
Gilbert, E.W. 'Andrew John Herbertson 1865-1915: an appreciation of his life and work', 313-31
Jay, L.J., 'A.J. Herbertson - his service to school geography', 350-61
Unstead, J.F., 'A.J. Herbertson and his work', 343-7

c. Herbertson Memorial Lectures
Fifteen of these lectures have been given since their foundation in 1917: the first thirteen of these are listed in *Geography*, vol 50, (1965), 371-2. Several of these contain biographical details about Herbertson, especially:
Rudmose Brown, R.N., 'Scotland and some trends in geography - John Murray, Patrick Geddes and

Andrew Herbertson', *Geogr.*, vol 33, (1948),
107-20

Fleure, H.J.,   'The later developments in Herbertson's
thought.   A study in the application of Darwin's
ideas', *Geogr.*, vol 37, (1952), 97-103

Myres, J.L.   'Region and race', *Geogr.*, vol 21,
(1936), 18-27

d. *Major natural regions*

A comparison of Herbertson's scheme with other
schemes is made in: 'Classification of regions of
the world', Report of a Committee of the Geographical
Associaton, *Geogr.*, vol 22, (1937), 253-82

Stamp, L.D.,   'Major natural regions - Herbertson
after fifty years' (Herbertson Memorial Lecture),
Geogr., vol 42, (1957), 201-16

Minshull, R.,   *Regional geography, theory and
practice*, London, 1967

2. *WORKS BY A.J. HERBERTSON*

A full list of the writings and editorial activities
of Herbertson is given in L.J. Jay, 'The published works
of A.J. Herbertson: a classified list', *Geogr.*, vol 50,
(1965), 364-70.

3. *UNPUBLISHED SOURCES*

Jones, G.B.,   'The life and career of A.J. Herbertson
and his contribution to the development of geography in Britain', M.A. (Educ.) thesis, University of Sheffield, 1959

Royal Geographical Society Correspondence Files: 104
letters written by A.J. Herbertson to the R.G.S.
between 1895 and 1915

In the possession of L.J. Jay: 26 letters written by
A.J. Herbertson to Hugh Robert Mill between 1893
and 1906, and one letter written by his wife
Dorothy Herbertson to H.R. Mill in 1906

*L.J. Jay is Senior Lecturer in Education in the
University of Sheffield.*

DATES	LIFE AND CAREER	ACTIVITIES, TRAVEL, FIELDWORK	PUBLICATIONS	CONTEMPORARY EVENTS
1865	Born 11 October at Galashiels, Selkirk-shire			
1872	Entered Galashiels Academy			
1879	To Edinburgh Institution			
1884				Scottish Geographical Society founded
1886	Student at Edinburgh University			
1889	Attended University of Freiburg	Living in Germany		
1891		Assistant Lecturer in summer school of Patrick Geddes (to 1899)		
1892	Awarded Physical Science Scholarship; became Demonstrator in Botany, University College of Dundee (summer term only)	Became Fellow of Royal Geographical Society; stayed at Ben Nevis Observatory, Scotland	*American life*	
1893	Married Frances L.D. Richardson	Carried out meteorological work in Paris and Montpellier		Geographical Association founded
1894	Lecturer, Owen's College, University of Manchester	Further work on meteorology in Scotland, at the Fort William observatory		
1895		Spent one term at Freiburg		6th International Geographical Congress, London
1896	Returned to Edinburgh to become Lecturer at Heriot-Watt College		Paper on education, given at 1895 Congress, published in the Report	
1897		For three months acted as Editor of *Scott. Geogr. Mag.*; Elected to the committee of the Geographical Association; Attended meeting of British Association in Toronto		
1898	Ph.D., University of Freiburg	Became Associate Editor, *J. School Geogr.*		
1899	Lecturer in Regional Geography, Oxford		*Man and his work* (textbook); some maps in Bartholomew's *Atlas of meteorology*	Oxford School of Geography founded
1900		Honorary Secretary, Geographical Association		

DATES	LIFE AND CAREER	ACTIVITIES, TRAVEL, FIELDWORK	PUBLICATIONS	CONTEMPORARY EVENTS
1901		Joint Editor, *Geogr. Teach.*	*Distribution of rainfall over the land* (thesis)	
1902	M.A., Oxford, by decree			
1904		Gave paper on 'Major natural regions' at Royal Geographical Society		
1905	Reader and Head of School Geography, Oxford, in succession to H.J. Mackinder	Recorder of Section E (*Geography*) British Association for the Advancement of Science	'The major natural regions' published in *Geogr. J.*	
1907		Member of the Committee on Water Supply	*Senior geography*	
1908	Elected Fellow of the Royal Meteorological Society			
1910	Professor, Oxford University	President, Section E (Geography), British Association for the Advancement of Science		
1912			'Thermal regions'	
1913			'The higher units'	
1914			*Oxford survey of the British Empire* published with Herbertson and O.J.R. Howarth as joint editors	Outbreak of war
1915	Died 31 July at Chinnor			

Philipp Melanchthon
1497–1560

MANFRED BÜTTNER

Photograph from the Archiv für Kunst und Geschichte

Philipp Melanchthon was a philosopher and theologian who received a classical education, a colleague and disciple of Martin Luther and the author of the first major work on Protestant dogma. He founded and organized the university and grammar school system in Protestant Germany and was influential in encouraging the study of geography.

1. EDUCATION, LIFE AND WORK

The son of an armourer, Georg Schwarzert, Philipp was born at Bretten in the Palatinate (Württemberg) and had one brother and two sisters. His unusual name comes from Johannes Reuchlin (1455–1522), his great-uncle, who turned it into its Greek form. He was brought up in a religious atmosphere and with his brother was educated for two years by Jakob Unger, a private tutor from Pforzheim engaged by his grandfather, Johannes Reuter. On Reuter's death in 1507 the whole family moved to Pforzheim and Philipp, already known to possess exceptional ability, went to the local grammar school, then renowned as second only to Schlettstadt in south-west Germany. With a number of other gifted pupils he attended the Greek classes given by Georg Simler, the director of the school and the author of one of the first Greek grammars to be published.

When Johannes Reuchlin, impressed by the ability of his great-nephew, took charge of his education he found that by the age of 12 Philipp could converse confidently with humanist scholars of the district. At the age of 13 Philipp went to Heidelberg University, where he was registered as a student on

14 October 1509. There, at the home of the influential Professor of Theology, Pallas Spangel, he met most of the leading humanists of the Oberrhein area. Among them was Helvetius Cunardus who, with Spangel, was a major influence on Philipp's development as a scholar. Cunardus, who lectured on astronomy and physics, had been a student of Caesarius of Julich, a mathematician, medical man and Greek scholar, and probably it was Cunardus who introduced Melanchthon to geography.

On 18 June 1571 Melanchthon received the first Bachelor of Arts degree but he was too young to study for a doctorate. On the advice of his great-uncle he moved to the University of Tübingen, a recent foundation of 1477, where the teaching was through the 'via moderna' which focused attention on Greek sources. Johannes Stöffler (1452–1531) became his main teacher, inspiring Melanchthon with a veneration for the classical sources of astronomy and geography. Stöffler was a highly critical scholar, always testing the veracity of his sources. He taught his students, among them Pellikan and Sebastian Münster, that science and especially geography led to the knowledge of God, and this view remained with Melanchthon throughout his life.

After Melanchthon received his Master's degree on 25 January 1514 he continued to work on theology and philosophy and turned also to law and medicine. He also lectured in the Faculty of Arts and Letters on Virgil, Cicero and Livius and taught Greek grammar. With this varied background he was becoming a universal scholar but in 1518, on the recommendation of Reuchlin to the Elector Friedrich of Saxony, he became lecturer in Greek and Hebrew at

the Universtity of Wittenberg.   In his inaugural
lecture, *De corrigendis adolescentiae studiis*, he
emphasized the close relation between humanism and
the Christian gospel.   Under Luther's influence he
qualified for the first degree of the Faculty of
Theology, the *Baccalaureus Biblicus*, conferred on 9 September 1519 and he remained a member of the two facul-
ties, Arts and Letters, and Theology, until his death.
His work on dogma and on the philosophy of Aristotle
earned him the title of *Praeceptor Germaniae*.   Some of
his hundreds of writings were repeatedly reprinted and
his school texts were used until the eighteenth century
in Protestant and even some Catholic schools.

Geography was not the principal concern of
Melanchthon, but he could not ignore a study that
to him was a convincing demonstration that God ruled
the world.   He edited the works of classical authors
with forewords and commentaries, including those of
Sacro Busto (*c.* 1230).   He wrote commentaries on
the works of his students, among them Caspar Peucer
(1525-1602), and influenced their work, as in the
case of Neander (1525-95).   He also published work
on physics, which would now be regarded as physical
geography.   Not all the published editions of his
widely circulated book, *Initia doctrinae physicae*,
have been found but it is known that there were
eight from Wittenberg between 1549 and 1587, one
from Frankfurt in 1550 and three from Leipzig in
1559, 1560 and 1563.   His geographical work, like
his theology, was intended for the universities.

As the *Praeceptor Germaniae* organizing the new
school and university system, Melanchthon encouraged
the teaching of geography.   He explained his
reasons for doing so in his theological works.
First, geography opened a *primum iter ad deum*, a first
way to God, though the greater second way came from
theology and the Bible.   A further value of geography lay in the light it shed on the scriptures.
Using books also available to and known by Sebastian
Münster a few years earlier, he accepted the view
that God must be at the centre of all geographical
study.   Stöffler at Tübingen probably instructed both
Münster and Melanchthon in basic geographical skills,
such as the Ptolemaic mathematical approach, while
Reisch was concerned with such problems as the
importance of geography for theology and the influence of Aristotle's philosophy.   Luther became
a close friend of Melanchthon and, in writing on the
doctrine of Providence in his main geographical
work, Melanchthon used Artistotelian concepts along
with the central Lutheran dogma of the constant
revelation of the divine power and purpose.

2. SCIENTIFIC IDEAS AND GEOGRAPHICAL THOUGHT

In 1531 Melanchthon published Sacro Busto's principal
work as a textbook for all the mathematical sciences
in Wittenberg with a foreword explaining the relation
between theology and science.   In the curricula for
Protestant schools and universities only subjects
that could serve the *Doctrina evangelica* were to be
taught.   The geography of the late medieval period
was based on the doctrine of creation and the work of
Ptolemy.   It showed what the world looked like,
through Ptolemy, and what happened at the Creation,

from the Bible and Aristotle.   But it did not reveal
the continuing purpose of God in the world, the
divine and merciful providence and mercy in human
affairs.   As Melanchthon explained in the *Initia
doctrinae physicae* of 1549, the work of the geographer should be extended to include physical and
human (cultural) geography on a religious basis.
Peucer studied the implications of this concept and
Neander worked on the definition of a *geographia
specialis* on a Lutheran foundation.

Melanchthon said that 'physics' should include
the quality and motion of natural bodies.   He recognized that there were 'natural laws', many of
which could only be explained by further research.
The study of 'physics' was in itself interesting,
useful in daily experience and a revelation of the
Christian doctrine of *providentia*.   In studying the
natural bodies, their movement and change, one
could proceed from cause to result, or from result
to cause, the latter being the more effective
approach.   Cosmology was concerned with the celestial and geography with the terrestrial sphere.
Nature, however, gave only a partial understanding
of God, made fuller by the doctrine of the Church.

Providentia, a major Lutheran concept, could
be understood in various ways.   Of these the first
was from observation of the existing world, the
'here and now' elements of life such as the year's
seasonal changes and the succession of day and night
together with the variation of weather and climate.
Melanchthon was concerned with aspects of physical
and human geography which could be observed day
by day, rather than with the past.

God is free, and not bound by his own laws as
many members of the Reformed Church had thought.
In a chapter on *Contingentia* Melanchthon emphasizes
that if God were not free, a scientist or theologian
might come to the idea of a sleeping God who retired
after the work of Creation.   In fact the Moravian
theologian, John Amos Comenius (1592-1670), came to
the view that God, having created the world, had
also determined human history in advance so that
his work was done. The Lutherans were not concerned
with a God of the past who supposedly had arranged
everything in advance but with a God currently at
work in the world, a merciful God caring for the
wellbeing of mankind at the present time and in
every place.   This view prevailed from the time
of A.H. Francke (1663-1727) to that of Kant.   Geography could give an understanding of the world
people knew and experienced, provided that it included not only the basic mathematical aspects but
also the physical aspects that affected the daily
lives of its inhabitants.

Having defined his outlook in the first part of
the *Initia*, Melanchthon provides in Part II a geography and cosmology on an Aristotelian plan.   He
begins with the Greek cosmos, observed to be finite
as the firmament moves during a measurable period
of time.   Clearly the earth must be round as most
of the visible stars move in a circular orbit and
no vacuum can exist.   He strongly denied the view
that the earth could revolve around the sun and
therefore the earth was the centre of the Cosmos and,
like the centre of any circular motion, must be in
a state of rest.   Any supposition that the earth is

not the centre of the Cosmos is contradicted by empirical observation, so land and water cover a spherical body with a circumference of 5,400 German miles (40,500 kilometres).   The geometrical centre of the world and the centre of gravity must be identical: since other hypothetical worlds must be ruled by the same rules as our own, and as no vacuum can exist, it is not possible that other worlds can exist and therefore the earth is the centre of the universe.

The continuing existence of the world has engaged the thought of men for millennia, and Aristotle has shown that natural scientific research might suggest permanence.   Melanchthon said that the Bible clearly explained the beginning of the world. Like others before him he considered that the heavens and the spheres existed in pure ether, which to him meant pure light.   The classicists thought that there were eight spheres but Melanchthon, and Reisch also, thought that there were ten.   Motion began in the extreme outer sphere and progressed towards the centre of the earth.   He rejected the idea that the stars influenced the world, but they were instead signs of God's influence.   God ruled the sun, the stars, the seasons and the climate, indeed everything on the earth including the air, all elements and every animated creature.   Everything existed for the salvation of mankind: everything was ruled from above, derived not from some kind of *causa* (cause) but from the God of the Bible for the good of man.   Such was the doctrine of *Providentia*.

Difficulties that arise in trying to prove God's rule of the world by means of classical physics were met in the chapters on *fatum* (fate).   Four kinds could be distinguished.   The first was God's deliberate decision; the second was the combination of natural causes; the third was the *fatum Stoicum*, the Stoic fate or combination of the first cause with all following causes; and the fourth was the *fatum Aristotelicum*, the order of natural causes. Melanchthon rejected totally the last two, for the *fatum Stoicum* limited the freedom of God and the *fatum Aristotelicum* could not envisage a deliberate decision of God.   The *fatum physicum* was nothing more than a revelation of how the world existed, and therefore the first kind of fate, deliberate decision, was another name for the providence of God.

Using inductive methods Melanchthon reached conclusions **in his earlier section about the** working out of God's purpose.   By noting the influence of the sun on weather, for example, Melanchthon worked on a deductive 'from below' basis derived from observations in cosmology and geography.   He showed that God ruled the rainfall and therefore all life depending on rainfall, so that rain and drought occurred at the right time. Natural laws, therefore, were not binding on the Deity, for they would preclude divine intervention in the life of the physical and human world.   The second, deductive, section gives a teleological approach. Melanchthon had originally intended to consider Plato's idea that God was the creator of the world, but turned instead to the idea of Aristotle that God was the living force of movement and change.   This led to a change in focus in geographical thought at the time of the Reformation through the theological emphasis on *Providentia* rather than *Creatio*.

The religious emphasis of Melanchthon and his disciples meant that the effect of Christianity on civilization became a part of the study of geography. Human, or cultural, geography was partly concerned with the distribution of religions.   Peucer in his textbook said that a Lutheran geographer must deal with the world in which God had placed mankind, and that the world was itself a series of divine revelations made understandable by the study of geographical phenomena.   Neander, a student of Melanchthon but also influenced by Peucer, wrote a *geographia specialis* based on the distribution of religion.

To summarize, Melanchthon's innovations in geographical thought were twofold.   First, he laid emphasis on the dogma of *Providentia* rather than *Creatio* as the source of understanding for geographers.   Second, having absorbed the mathematical approach of Ptolemy into a broader physical and cultural geography, he was the first German scholar to develop the scientific basis of *Vollgeographie*, a complete geography, that has survived, however modified, to the present day.

3. INFLUENCE AND SPREAD OF IDEAS

Melanchthon's great achievement was to institute the inclusion of nature and man in geographical studies. Cosmographical and geographical facts were interpreted in a teleological manner.   His view that geography had to take account of the spread of Christianity led to study of the distribution of religions all over the world.   The emphasis on *Providentia* in relation to geography was a contribution to the thought of the Reformation, seen for example in the work of Münster.   Keckermann, the founder of scientific geography, owed much to the influence of Melanchthon.   Kant's writing also shows the influence of Melanchthon, especially in his treatment of man in the *Physical geography*. There are some traces of Melanchthon's views in the work of Carl Ritter, though later he adopted a transcendental philosophy which ruled out the idea that proofs of God's existence may be found in geographical facts.   In time the views of Melanchthon became unacceptable, but his broadening vision of geography was for a time an inspiration to later scholars.

Bibliography and Sources

1. REFERENCES ON MELANCHTHON
a. General
Bergmann, H., 'Philipp Melanchthons Ansichten von dem Wert und der Bedeutung der einzelnen Unterrichtsgegenstände' ('Melanchthon's opinion on the value

and importance of the different subjects of instruction'), *Rheinische Bl. Erzich. Unterricht.* (1897), 341-55

Bernhard, W., *Philipp Melanchthon als Mathematiker und Physiker (Melanchthon as mathematician and physical scientist)*, Wittenberg (1865)

Blumenberg, H., *Die kopernikanische Wende (The Copernical change)*, Frankfurt a.M. (1965), 100-21

Bornkamm, H., *Luther und das Naturbild der Neuzeit (Luther and the view of Nature in modern times)*, Berlin (1937)

Geyer, H.G., 'Mensch und Welt zum Problem des Aristotelismus bei Melanchthon' ('Man and the world - about Aristotelism in Melanchthon's work'), Doctorial thesis, Bonn (1956)

Günther, S., *Geschichte des mathematischen Unterrichts im deutschen Mittelalter (The history of mathematical instruction during the Middle Ages in Germany)*, Berlin (1887)

Hammer, Wilh., *Die Melanchthonforschung im Wandel der Jahrhunderte (Research on Melanchthon in the different centuries)*, vol I, 1519-1799, vol II, 1800-1965. Reprinted in *Quell. Forsch. Reformationsgeschichte*, vols XXXV/XXXVI, Gütersloh (1967-8)

Hartfelder, K., 'Philipp Melanchthon als Praeceptor Germaniae' ('Melanchthon as the Praeceptor Germaniae'), *Monumenta Germaniae Pedagogica*, vol VII, Berlin (1889)

Hofmann, Fr., 'Philipp Melanchthon und die zentralen Bildungs probleme des Reformationsjahrhunderts. Ein Beitrag zur erziehungsgeschichtlichen Wertung des 16. Jahrhunderts' ('Melanchthon and the central questions of education during the century of the Reformation. A contribution to the evaluation of the 16th century for the history of education.) In *Philipp Melanchthon 1497-1560. Vol. 1 Philipp Melanchthon Humanist, Reformator, Praeceptor Germaniae*, Berlin (1963)

Maurer, W., 'Melanchthon und die Naturwissenschaften seiner Zeit' ('Melanchthon and the physical sciences of his times'), *Arch. Kult-gesch.*, eds Wagner, Borst, Grundmann, vol 44, Köln/Graz (1962)

Petersen, P., *Geschichte der aristotelischen Philosophie im protestantischen Deutschland (History of the Aristotelian philosophy in Protestant Germany)*, Leipzig (1921)

Wagenmann, A.J., 'Melanchthon, Philipp', *Allg. Dtsch. Biogr.*, vol 21, (1885), 268-79

b. The geographer

Bonacker, W. and Volz, H., 'Eine Wittenberger Weltkarte aus dem Jahre 1529' ('A world-map of the year 1529 made in Wittenberg), *Die Erde Z. Gesell. Erdk.*, vol 87, (1956), 154-70

Büttner, M., 'Theologie und Klimatologie' ('Theology and Climatology'), *Neue Z. für Syst. Theologie und Religionsphilos.*

Büttner, M., 'Geographie und Theologie im 18. Jahrhundert ('Geography and theology during the 18th century'), *Vehr. Dtsch. Geogr. 1965*, Bochum Wiesbaden, (1966), 352-9

Büttner, M., 'Zum Gegenüber von Naturwissenschaft (insbesondere Geographie) und Theologie im 18. Jahrhundert. Der Kampf um die Providentialehre innerhalb des Wolffschen Streits' ('The relation between science and theology during the 18th century. The controversy on the doctrine of providence with reference to the views of Wolff'), *Philos. Nat.*, vol 14, (1973), 95-122

Büttner, M., 'Religion and geography. Impulses for a new dialogue between Religionswissenschaftlern and geographers', *Numen*, vol 21, Leiden, (1974), 163-96

Büttner, M., 'Regiert Gott die Welt? Vorsehung Gottes und Geographie. Studien zur Providentialehre bei Zwingli und Melanchthon') ('Does God rule the world? God's providence and geography. Studies about the doctrine of providence in the work of Zwingli and Melanchthon'), *Calwer Theologische Monographien*, vol 3, Stuttgart (1975)

Büttner, M., 'Die Emanzipation der Geographie im 17. Jahrhundert' ('The emancipation of geography during the 17th century'), *Sudhoffs Arch.* vol 26, (1975), 1-16

Büttner, M., 'Die Bedeutung der Reformation für die Neuausrichtung der Geographie im protestantischen Europa' ('The significance of the Reformation for the new orientation of geography in Protestant Europe'), *Archive für Reformationsgeschichte* (forthcoming)

Büttner, M., 'Von der Religionsgeographie zur Geographie der Geisteshaltung' ('From the geography of religion to the geography of the mental attitude'), *Die Erde Z. Gesell. Erdk.*, vol 107, (1976), 300-29

Müller, K., 'Philipp Melanchthon und das kopernikanische Weltsystem' ('Melanchthon and the systematic world picture of Copernikus'), *Centaurus. International Magazine of the History of Mathematics, Science, and Theology*, vol 9/1, Copenhagen, (1963), 16-28

2. SOURCES

1519 *De corrigendis adolescentiae studiis*, Wittenberg, rpt *Corpus Reformatorum I*, 52p.

1531 *Johannis de Sacro Busto libellus de sphaera*, Wittenberg (ed.) rpt (foreword) *Corpus Reformatorum II*, 530p.

1536 *De astronomia et geographia*, Wittenberg, rpt *Corpus Reformatorum II*, 292p.

1549 *Initia doctrinae physicae, dictata in Academia Wittenbergensi*, Wittenberg, rpt *Corpus Reformatorum XII*, 1p. *Encomium Franciae*, rpt *Corp. Reform. XI*, 383p. *Encomium Sueviae*, rpt *Corp. Reform. XI*, 374p. *Declamatio de Misnia*, rpt *Corp. Reform. XII*, 34p. *De Salinis Saxonis*, rpt *Corp. Reform. XII*, 119p. *Explicatio locorum Palaestinae*, rpt *Corp. Reform. XX*, 439p.

521/1559 *Loci*, Engelland, H.V., ed Gütersloh, 1952

Manfred Büttner, Dr. phil., Dr. rer.nat., Dr. theol., is Professor of the History of Geography and of Cultural Geography at the Ruhr Universtüt, Bochum.

DATES	LIFE AND CAREER	ACTIVITIES, TRAVEL, FIELDWORK	PUBLICATIONS	CONTEMPORARY EVENTS AND PUBLICATIONS
1497	Born 16 February in Bretten, Württemberg			
1499				A. Vespucci's first voyage
1504				*Cosmographiae introductio*, St Dié
1507	Death of his father; attends the *Latein schule* at Pforzheim			
1509	Student at Heidelberg University			
1511	Bachelor of Arts			Stöffler professor of mathematics in Tubingen
1512	Continues studies in Tübingen			
1512/14				Stöffler's lectures on Ptolemy's geography
1514	*Magister artium*			
1517				Beginning of the Reformation in Germany
1518	Inaugural lecture at Wittenberg			
1519	Baccalaureus Biblicus; member of the Faculty of Theology			
1521			First edition of the *Loci communes*	
1528		New organization of institutional and ecclesiastical affairs in Saxony		Münster organizes a staff to describe Germany and Europe. Stöffler's death
1531			Edition of Sacro Busto's *De sphaera*	
1546				Luther's death
1549		Foundation of Lutheran geography	*Initia doctrinae physicae*	
1550				Final edition of Münster' *Cosmographia*
1560	Died in Wittenberg, 19 April			

Sebastian Münster
1488–1552

Title page of Cosmographie

MANFRED BÜTTNER AND
KARL H. BURMEISTER

Sebastian Münster had a classical education with an emphasis on Hebrew, and later studied oriental languages. He spent 24 years as a member of the Franciscan Order, but in 1529 joined the Swiss Reformed Church. His publication of the first complete Bible in Hebrew made him famous but in his later years he became renowned as a geographer. Melanchthon had given a scientific basis to post-Reformation geography but Münster, though strongly influenced by Melanchthon, added a consideration of what became anthropogeography, that is, human or cultural geography. Like Melanchthon he was concerned with cosmography, which he defined as a 'description of all countries, territories, and towns etc. of the whole world'. Theoretically he made a distinction between cosmography as the description of the whole world and geography as the description of the earth's surface: in fact his work is mainly on geography. Only a few pages of his famous *Cosmography* of 1544 are given to that subject and the remainder deal with geography, with a strong human interest that contrasts with the dominant physical emphasis of Melanchthon. The emphasis is on the earth and man, on *Länderkunde*. This connection between the earth and man was not an aspect of Greek classical geography and in the work of Münster it was derived from the idea of *Providentia*, which in the Swiss Reformed Church became steadily more human in conception.

1. EDUCATION, LIFE AND WORK

Münster was born at Niederingelheim, near Mainz, but little is known of his family background except that he had at least one brother whose son studied at various European universities. Münster appears to have known this nephew well and to have praised his work for the German Reformation. Apparently Münster's family belonged to the cultured middle class of the day and Sebastian received the usual education, the *Trivium*, or training in the liberal arts, given to pupils who were to enter the universities, and as no local school provided a suitable course he was taught mainly by clergy. Together with the religious atmosphere of his home this gave him a lifelong interest in theology.

In 1505 he entered the Franciscan Order at Heidelberg and began the next stage of preparation for the university, the *Quadrivium*. His studies included logic, ethics and metaphysics, as well as cosmology. Two years later, in 1507, he went to Louvain to study mathematics, astronomy and geography but he is thought to have stayed there only a short time and to have moved to Freiburg as a student of Gregorius Reisch (1467-1525), a theologian, Hebrew scholar and geographer. The remainder of Münster's life was spent on these three subjects and he appears to have been inspired by the teaching of Reisch. In 1509 the Franciscan Order sent him to Rufach to study the same subjects with Conrad Pellikan (1470-1556). He was Pellikan's favourite student and by 1510 had compiled a Hebrew grammar and dictionary. In time his interest in geography became so strong that Pellikan warned him not to neglect theology and philosophy. In 1511 Münster joined Pellikan at the Franciscan monastery in Pforzheim and in 1512 he was ordained priest.

By 1514 Pellikan was at Tübingen and Münster became a lecturer in theology and philosophy there. At Tübingen he encountered Johannes Stöffler (1452-1531), with whom he studied geography. Münster's biographer, Schreckenfuchs, said that he was drawn to Tübingen by Stöffler rather than by Pellikan. Once again Münster became a favourite student with full access to the papers of his patron. Most of these were destroyed in a fire, but excerpts were in Münster's possession. At Tübingen he also met Melanchthon, then a student of Stöffler.

In 1518 Münster moved to Basel and worked with the printer Adam Petri. The close link with Pellikan was now renewed and his work was once more directed towards theology. Pellikan arrived in Basel in 1519 and published Martin Luther's complete works in the same year. Münster worked closely with Pellikan, and in 1520 was responsible for the translation of Luther's *Decem praecepta Wittenbergensi praedicta populo*. Already Münster was well versed in the theology of the Reformation, but his public profession of acceptance was delayed until 1529.

Meanwhile, in 1520, he published a Hebrew grammar and became a lecturer at Heidelberg University. He became Professor of Hebrew in 1524 and extended his linguistic researches and publications to Aramaic, and in time also to Arabic and Ethiopian. His interest in cosmography was revived by friendship with Beatus Rhenanus (1485-1547). On his move to the University of Basel as Professor of Hebrew in 1529 he married the widow of Adam Petri, whose son, Heinrich, published much of his work. From 1542-4 he was Professor of Old Testament Theology at Basel, and from 1547-8 Rector of the university. His teaching of Hebrew at Basel brought widespread renown to the university and one of the students who achieved lasting fame was Jean Calvin.

Less than a quarter of Münster's known publications deal with geography or cosmography. He never taught geography, but he regarded theology and geography as a unity. Theology, especially of the Old Testament, with Hebrew, showed the *secundum iter ad deum*, the second way to God. Geography, within the framework of natural theology, *theologia naturalis*, pointed the first way, the *primum iter ad deum*, the first or lesser vague approach as a basis for the later clear and shining understanding given by the Bible. Why Münster never taught geography has not been explained, for other scholars such as Vincentius (*c*. 1190-1264) and Reisch, his contemporary, did so.

Geographical work by Münster included the publication of the books of such classified geographers as Ptolemy, Mela and Solinus, with commentaries. He also published original works, culminating in his famous *Cosmography* of 1544. All his earlier works had been by way of preparation for it. They included the *Germania descriptio* of 1530, the *Mappa Europae* of 1536 and his commentary on Ptolemy, published in four editions between 1540 and 1552 and translated into Italian in 1548. The *Cosmography*, published by Petri in Basel, went into twenty-one editions between 1544 and 1628, and the Latin edition appeared five times. In its day it was one of the most widely read books. Six editions appeared in French, five in Italian and one in

Czech (1554). As the initial work of 1544 was unfinished, it is likely that the 1550 issue is the final version, though in later editions historical information is brought up to date. The *Cosmography* was apparently intended for the educated general public and therefore published in modern languages rather than in Latin: even the illiterate could learn something from its abundance of illustrations. The whole conception of Münster's geographical writing was that the environment was part of God's creation, itself leading people to the central Christian truth proclaimed in the Bible.

During his time as a Franciscan at the universities of Heidelberg, Louvain and Freiburg Münster knew Reisch, Pellikan and Stöffler, all scholars with geographical interests, and it is relevant to ask how each influenced him. Probably Reisch showed the apparent divergence between the Bible and the thought of the classical Greek geographers as a philosophical problem, while Stöffler gave Münster instruction in the basic geographical skills including the mathematical approach of Ptolemy and the topographical work of Strabo and Mela. It is thought today that Pellikan's influence on Münster was of only minor significance and that that of Stöffler, who could be regarded as the first professor of geography in Germany, was far more significant. Melanchthon was to be a crucial influence later, for he gave Münster a philosophical approach with a definite Protestant orientation.

However the change of religious allegiance was actuated, it is clear that Münster came to value the outlook of Martin Luther and his disciples, though he was also interested in the philosophy of Zwingli. Münster gave no attention in his writing to Aristotle, but perhaps this showed his lack of enthusiasm for philosophy. In his cosmography he was helped by the approach of Rhenanus, who encouraged him to publish a book in 1525 which included a map of Germany. Rhenanus does not appear to have contributed to the philosophical outlook of Münster, who, as a Hebrew scholar, increasingly appreciated that the Bible, and especially the Old Testament, could not be understood without a knowledge of geography. To Münster classical Greek geography was not an end but only a beginning, and the need of his own time was to find a new geography, a cosmography, incorporating his own observations and those of his colleagues, in tune with the new thinking of the Reformation.

2. *SCIENTIFIC IDEAS AND GEOGRAPHICAL THOUGHT*

In the foreword to his main geographical works Münster explains his views on the task, aims and methods of geography. He makes three points. First, geography sheds light on the Bible, for anything that happens on earth is in accordance with the will and providence of God. Second, geography gives ordinary people who are unable to travel widely a realization of the creation and rule of the world by God. This may be done by giving information on the mountains, rivers, lakes, peoples, landscapes, cities, principalities, religions, and much besides of all parts of the world, especially the homeland of Germany. Third, these additions to the knowledge of men must be introduced by a study of cosmography as a science.

In Part I of his work Münster began with the Creation, but in his discussion he differed from Catholic theologians, such as Vincentius, who were particularly concerned with what happened on the first two days of the account given in the book of Genesis. Münster was especially interested in what happened from the third day, when the water was divided from the land. This was to him, and to other Reformation scholars, the time when divine providence began: he was not concerned with the earlier planning of the world.

The centre of interest for Münster became the surface of the earth, divided between land and water, which he believed was still as it had been on the third day of Creation. He was the first scholar to discard the cosmography of Aristotle, who worked from the outer to the inner spheres of the universe. In fact, he was seeking a compromise between Greek thought, the biblical account of creation and the theology of the Reformation by directing attention to the surface of the earth. The earth was created for the good of mankind and God continued to rule and modify the natural world, in which islands might come into existence and plants would provide food. Ten chapters of his *Cosmography* deal with the theologically conceived world, noting that the three upper elements had been created in the form of a hollow sphere. He added a short treatment of the geography of Ptolemy.

In dealing, briefly, with settlements and population Münster went back only to the Flood, and not to the Garden of Eden. He explained that all settlements were destroyed in the Flood, but that the few people who survived and settled in Armenia were the founders of all the new settlements and populations. He gave a summary treatment of the great empires with their rise and fall, and noted that Germany arose through the destruction of the Roman Empire. This completed Part I, the 'general' geography.

Part II, which may be regarded as a 'special' geography, was divided into six books, of which the first was a foreword in which Münster explained his intention of dealing with individual countries. Following the practice of Ptolemy he proceeded from west to east. The second book of Part II dealt with Europe as a whole, followed by England, Spain, France and Italy. In the third Münster gave a general survey of the German people, including their distribution and religious allegiance, and then a detailed study of the Rhine from Switzerland northwards, followed by Swabia, Bavaria and Austria, and finally the northern lands from Hesse, Pomerania and Silesia. The fourth book dealt with the northern countries, including Denmark and Sweden, and those of the east with Hungary and Poland. Book five was on Asia, with special reference to Palestine, and mention was made of America as the 'New India' after the description of China. In the last book North Africa and Ethiopia were treated with a paragraph on Ptolemy after the treatment of Alexandria.

In presenting his material Münster usually worked from the general to the particular. The general aspects of a continent included its size, shape, fertility, climate, population, mountains and rivers. This was the normal practice of Ptolemy. Then came the special aspects, as they did in Ptolemy's work. Separate areas might be political entities, such as kingdoms and principalities, areas with natural boundaries, such as islands and peninsulas, or ethnic or religious units, such as the *Germania* of the German people or the African Christian empire of Prester John. The conformity with Ptolemy was most clear in Münster's treatment of areas possessing natural boundaries, but he followed others in using a variety of criteria, of a social, religious and political character, to define areas.

The varied approach in Münster's *Cosmography* is seen in his treatment of the British Isles and Germany. Ptolemy had dealt separately with England, Scotland and Ireland but in Münster's work these three countries are regarded as one unit. He wrote only a short treatment of the physical features, but much more on the political and religious history, the division into dioceses, the wars between aspiring and conflicting kingdoms, and customs, languages, art and science. Germany was conceived by Münster as an ethnic rather than physical unit. He opened his discussion at some length with the migrations of the German people, and stressed that any area inhabited or ruled by Germans was Germany. Holding the view that Germany had emerged as the successor to the Roman Empire, he dealt with the ecclesiastical and political history from the fall of that Empire, including some data on physical geography, such as climate, rivers, lakes, soil fertility. In the successive editions of his work from 1550, this book was gradually expanded from 100 to 200 pages. It ended with a paragraph on customs, and a list of the most important cities. He followed the same lines in the treatment of the various German kingdoms and principalities ruled by prince-bishops and other nobility. In the third book of Part II he discussed Helvetia (eventually Switzerland), whose people were of Celtic stock, living between the German lands on the north, Italy on the south and France on the west. He considered the mountains, rivers, climate, soil fertility and other physical aspects as well as the cities which had an interesting religious, cultural and political history and were significant cultural and trading centres. In the *Cosmography*, Münster made a substantial contribution to the knowledge of the German lands, on which little had been said by Ptolemy. However, much of the content had been provided by others, so to some extent the *Cosmography* is a work of compilation.

There are three maps in Part I of Münster's *Cosmography*, here discussed in terms of their geographical conceptions rather than for their cartographical qualities. The first map shows the whole world as Münster conceived it to be on the third day of Creation, with a clear division of land and water, the elements of air and fire above the surface, the stellar space having fixed stars and the celestial sphere. A sailing boat is seen on the water and animals on the land, so in fact it represents the time after the expulsion from Paradise.

The map as a whole shows a combination of biblical and classical Greek thought on the nature of the earth.   The second map includes the three known parts of the earth's surface, Europe, Asia and Africa, but also includes and names America, which figures in Münster's text as 'New India'.   The third map is a reproduction of Ptolemy's famous representation of the surface of the earth, which was a basic source for the text.

The maps in the 'special' geography raise some interesting points.   Some are oriented from south to north, notably those for parts of Germany, and some maps are oriented from west to east, such as those for the British Isles.   With each new edition the orientation from north to south became more common.   Münster did not always manage to transfer his knowledge from the text to the map. For example, from his own researches on the Rhine near Speyer he knew that the river had a marked curve to the east, but this does not appear on the map.   He was a pioneer in making city maps and included views of cities as illustrations, but it is doubtful if all are authentic.   In the 1580 edition there are maps of American cities which must have been added by other hands after his death in 1552.

Münster used all available written sources and supplemented them by his own travels and those of his friends and disciples.   Under his inspiration a number of people worked on the geography of the upper Rhine.   Before his time geographers generally derived their material from individual scholars as Cochlaus did from Mela, and Stöffler from Ptolemy.   But Münster showed breadth of learning both in his 'general' and 'special' geography.   Although Reisch and Vincentius had tried hard to find a compromise between the biblical doctrine of creation and classical geographical thought, Münster went further and looked for a compromise between the central Protestant dogma of the divine rule of the world and classical Greek geographical thought.   Never before had general geography been concerned with the human aspects of geography and undoubtedly this development was inspired by Melanchthon, who said that geography must be in accord with *Doctrina Evangelica*. Mathematical geography in itself did not prove and demonstrate the divine rule of the world, and its maps only indicated the world's appearance.   Geography must therefore be a study of broader range, including man and his history.

Division into 'general' and 'special' geography was practised by others before Münster, among them Glaeren.   The originality of Münster lies in the stress given to 'special' geography and the revelation of the increasingly detailed study that could be followed, for example in the work on Germany. Münster was influenced by previous workers, especially in the arrangement of his material.   He followed Ptolemy in dealing with large areas, and other predecessors, including Strabo and Mela, in dealing historically with individual countries.   In his work on Germany he used both the east-west sequence of Ptolemy and the north-south sequence of Cochlaus.

Another interesting aspect of Münster's 'special' geography is his flexibility of approach.   For areas in which it appeared to be relevant the physical geography is stressed, as in the mountainous character of Switzerland or in the climatic relation to soil fertility in Europe and Africa.   Here we can see an acute awareness of population and transport as human distribution.   Subtlety of approach is also seen in his acceptance of ethnic, religious, political or cultural boundaries where they seemed to be more significant than natural boundaries.   He makes advances also in urban geography, for though Cochlaus was content just to mention Nürnberg as the principal city of Germany, Münster included all the cities, noting their geographical locality, history and significance for the surrounding area.

In his own time there was some criticism of Münster's work.   This arose partly from his use of the terms cosmology, the theory of the universe as an ordered whole, cosmography, the science which describes the general features of the universe (the heavens and the earth), and geography.   Until the early sixteenth century geography and cosmography had generally been regarded as one and the same study, and their separate distinction was to be a problem for post-Reformation geographers.   In his work *Cosmography or the description of all countries, principalities...on the surface of the earth* Münster actually provided a mathematical description of the world according to Ptolemy and a description of the natural and historical aspects of the earth -- that is, physical and human geography.   Only in Part I, the general section, does he fulfil the promise of shedding light on the secrets of the Bible and of God's purpose in the world. Also, Part I deals with the entire world and not with its parts as his title would suggest.

The main theological purpose scarcely emerges in the second part of his work, the 'special' geography, and in his descriptions of individual countries he includes some rather unconvincing and obscure detail which may have been a concession to the taste and interests of contemporary readers, perhaps influenced by the wishes of his son-in-law who published his works and naturally wanted a 'best-seller'.   Nevertheless Münster, influenced by Protestant thought and especially by the *Doctrina Evangelica* of Melanchthon, and well versed in Greek geographical ideas, looked towards a 'complete' geography, including the mathematical, physical and cultural aspects of the surface of the earth and the life of its inhabitants. This broad conception of the subject opened ways to a far wider range of studies than Münster, or any of his contemporaries, could have foreseen.

3. INFLUENCE AND SPREAD OF IDEAS
Since Münster had no students, his influence was spread solely through his books.   In the seventeenth century Mercator used Münster's work, though his main concern was to develop cartography.   Keckermann, founder of the *geographia generalis*, moved towards mathematical geography and away from description of individual countries.   Later, in the work of Varenius, the emphasis was on physical geography. But Kant followed Münster's practice of avoiding any rigid model and stressing the characteristic and prominent (and even the curious) features of a country, though he took most of his material from

Busching (1724-93), whose general conception of geography differed from that of Münster and whose work superseded the *Cosmography* of Münster.

To a great extent the fascination of Münster's work lies in the relation of his geography to the theological and philosophical outlook of his time, and his influence on those who were concerned to place geography in a universal system of thought was considerable. That the influence of Münster's thought may appear to be ephemeral is partly due to its emergence at the time when geographers were labelled by religious allegiance, Stöffler as a Catholic, Melanchthon as a Lutheran and Münster as a member of the Swiss Reformed Church attracted by the outlook of Zwingli. Of permanent significance is Münster's preoccupation with individual countries, in contrast to the general geography of the Lutheran geographers, notably Melanchthon, Peucer, Varenius and Kant.

Bibliography and Sources

REFERENCES ON SEBASTIAN MÜNSTER

a. General
Burmeister, K.H., 'Neue Forschung zu Sebastian Münster ('Recent research on Sebastian Münster'), *Beitr. Ingelheimer Gesch.*, vol 21, Ingelheim (1971), 42-57. (A bibliography containing almost 300 items on Münster is included in this paper.)

b. Biographies
Burmeister, K.H., 'Sebastian Münster. Versuch eines biographischen Gesamtbilds' ('Sebastian Münster. An attempt at an inclusive biography'), *Basler Beitr. Gesch.-Wiss.*, vol 91, Basel/Stuttgart (1963), 2nd ed.
Burmeister, K.H., ed, *Briefe Sebastian Münster, Lateinisch und Deutsch (Münster's letters in Latin and German)*, Frankfurt, 1964
Hantzsch, Victor, 'Sebastian Münster. Leben, Werk, wissenschaftliche Bedeutung' ('Sebastian Münster. His life, work and scientific importance'), *Abh. K. Sächsischen Akad. Wiss., Philol. Hist. Kl.* vol 18, no 3, Leipzig, (1898), 188 p.

c. Further writings on Münster
Buczek, K., 'Ein Beitrag zur Enstehungsgeschichte der Kosmographie von Sebastian Münster' ('On the history and background which influenced the creation of Münster's cosmography'), *Imago Mundi*, vol 1, (1935), 35-40
Büttner, M., *Die geographia generalis vor Varenius. Geographisches Weltbild und Providentialehre (The geographia generalis before Varenius. The geographical world picture and the Doctrine of Providence)*, Wiesbaden, 1973

Büttner, M., 'Die wechselseitige Beziehung zwischen Weltbild und Glaube vom Mittelalter bis zur Neuzeit' ('The interrelation between the world view and faith from the Middle Ages up to modern times'), *Weltbild und Glauben*, Evangelische Akademie Baden, ed Baden, Bad Herrenalb (1976), 30-74
Büttner, M., 'Die Bedeutung der Reformation für die Neuausrichtung der Geographie im protestantischen Europa' ('The significance of the reformation for the reorientation of geography in Protestant Europe'), *Archive für Reformationsgeschichte*, vol 68, (1977), 209-25
Gattlen, A., 'Zur Geschichte der ältesten Walliserkarte' ('On the history of the oldest map of Vallis'), *Vallesia*, vol 8, Sitten, (1953), 101-20
Grenacher, F., 'Die erste Rheinstromkarte im 16. Jahrhundert geschaffen' ('The first map of the Rhine river made in the 16th century'), *Strom und See*, Basel, (1956), 452-5
Grenacher, F., 'The Basel proofs of seven printed Ptolemaic maps', *Imago Mundi*, vol 13, (1956), 166-71
Grenacher, F., 'Die älteste Landtafel der Regio Basiliensis' ('The oldest map of the Basel region'), *Regio Basiliensis*, vol 1, (1968), 67-85
Horch, H.J.W., 'Sebastian Münster: Mappa Europae', *Rev. Hist.*, vol 87, (1971), 187-220
Horch, H.J.W., 'Bibliographische Notizen zu Sebastian Münsters Baseler Ausgaben der "Geographia universalis" des Ptolemaeus' ('Bibliographical notes on Münster's edition of Ptolemey's "Geographia generalis"'), *Gutenberg-Jahrb.*, (1973), 257-66
Horch, H.J.W., 'Bibliographische Notizen zu einigen Ausgaben der *Kosmographie* von Sebastian Münster und ihre Varianten' ('Bibliographical notes on some of Münster's editions of his *Cosmography* and its variations'), *Gutenberg-Jahrb.*, (1974), 139-51
Jenny, B.R., 'Zu Sebastian Münster' ('On Sebastian Münster'), *Schweiz. Z. Gesch.*, vol 15, (1965), 87-97
Knapp, M., *Zu Sebastian Münsters 'astronomischen Instrumenten' (On Münster's 'astronomische Instrumente')*, Basel, 1920
Koncyñska, W., *List Sebastian Münstea do Stanislawa Laskiego i garsc szczegolow w zwiazko z jego Kosmografia (Münster's letter to Stanislaus Laski on some characteristics of the Cosmography)*, Cracow, 1935
Matthey, W., 'Sebastian Münsters Deutschlandkarte von 1525 aus einem Messingastrolabium' ('Münster's map of Germany, 1525'), *Jahresber. Ger. Natl.-Mus.*, vol 106, (1951), 42-52
Oehme, R., 'Sebastian Münster und die Donauquellen' ('Münster and the sources of the river Danube'), *Alemannisches Jahrbuch*, (1957), 159-65
Panzer, W., *Der deutsche Geograph Sebastian Münster (Münster, a German geographer)*, Ingelheim, 1953
Ruland, H.L., 'A survey of the double-page maps in 35 editions of Münster's *Cosmographia Universalis* 1544-1628 and his edition of Ptolemy's *Geographia* 1540-1552, *Imago Mundi*, vol 16 (1962),

84-97

Siegrist, W., 'A map of Allgäu, 1534', *Imago Mundi*, vol 6, (1950), 27-30

Schilling, F., 'Sebastian Münsters Karte des Hegaus und Schwarzwalds von 1537' ('Münster's map of the Hegau and the Black Forest, 1537'), *Jahrbuch der Coburger Landesstiftung*, (1961), 117-38

Wilsdorf, H., 'Präludien zu Agricola. Die Bergbaukunde und ihre Nachbargebiete in der Cosmographey des Sebastian Münster' ('Preludes to Agricola. On mining and neighbouring subjects in Münster's Cosmography'), *Freiburger Forschungshefte*, Berlin, (1954), 65-205

Wolkenhauer, A., 'Sebastian Münsters handschriftliches Kollegienbuch aus den Jahren 1515-18 und seine Karten, Abhandlungen der Gesellschaft der Wissenschaften zu Göttingen' ('Münster's college book in manuscript for the years 1515-18 and his maps and essays at the Society of Science at Göttingen'), *Philol.-Histor. Kl.* vol 11, no 3, Berlin 1909

2. SELECTIVE BIBLIOGRAPHY OF WORKS BY SEBASTIAN MUNSTER

a. General:
Burmeister, K.H., *Sebastian Münster. Eine Bibliographie*, Wiesbaden, 1964. (This bibliography mentions all the works Münster wrote, published or translated, in all 174 items.)

b. Geographical work
1530 *Germaniae descriptio*, Basel
1532 *Weltkarte su Simon Grynaeus. Novus orbis*, Basel
1536 *Mappa Europae*, Frankfurt, 2nd. ed., 1537
1538 *Edition and translation into Latin of Aegidius Tschudis Rhaetia*, Basel
1538 Edition of *Solin and Mela*, Basel
1540 Edition and translation into Latin of *Geographia universalis Claudii Ptolemaei*, Basel, 4th. ed., 1552. (Italian ed., Venice 1548)
1544 *Cosmographia*, small German ed., Basel, 4th ed., 1544
1550 *Cosmographia*, large German ed., Basel, 17th ed., 1628; Latin ed., Basel, 1550, 5th ed., 1574; French ed., Basel, 1552, Paris, 1575; Czech ed., Prague, 1554; Italian ed., Basel 1558, Venice 1575

c. Facsimile editions
1968 Oehme, Ruthardt, ed., *Cosmographia*, (large German ed., Basel 1550), Amsterdam
1968 *Sphaera Mundi des Abraham bar Chija*, new ed., Amsterdam
1965 Stopp, Klaus, ed., *Mappa Europae*, Wiesbaden

d. Astronomical and mathematical works
1525 *Instrument der Sonnen*, Oppenheim
1528 *Erklerung des Instruments der Sunnen*, Oppenheim
1529 *Instrument über den Moonslauff*, Worms
1529 *Erklerung des Instruments uber den Mon*, Worms
1534 *Instrumentum novum*, Basel
1534 *Canones super novum instrumentum luminarium*, Basel
1536 *Organum uranicum*, Basel
1526-8 *Kalender auf die Jahre 1527, 1533, 1549*, Basel
1531 *Compositio horologiorum*, Basel
1546 Edition of *Sphaera Mundi by Abraham bar Chija*, Basel
1551 *Rudimenta Mathematica*, Basel

e. Works in Hebrew
1520 *Epitome hebraicae grammaticae*, Basel
1523 *Dictionarium Hebraicum*, Basel, 6th ed. 1564
1524 *Institutiones grammaticae in Hebraeam linguam*, Basel
1525 *Grammatica Hebraica absolutissima Eliae Levitae*, Basel, 5th ed., 1552
1525 *Composita verborum et nominum Hebraicorum Elia Levita autore editum*, Basel, 2nd. ed. 1536
1527 *Chaldaica grammatica*, Basel
1527 *Dictionarium chaldaicum*, Basel
1527 *Capitula cantici autore Elia Levita*, Basel
1527 *Compendium Hebraicae grammaticae*, Basel, Paris, 3rd. ed., 1537
1530 *Dictionarium trilingue*, Basel, 3rd. ed., 1562
1535 *Isagoge elementalis in Hebraicam linguam*, Basel, 2nd. ed. 1540
1539 *Accentuum Hebraicorum liber ab Elia Iudaeo editus*, Basel
1542 *Opus grammaticum consummatum*, Basel, 6th. ed., 1570. The large number of his Hebrew-Latin editions include 2nd. ed., 1546
1534-5 *Hebraica Biblia Latina*, 2 vols, Basel, 2nd. ed., 1546
1537 *Evangelium secundum Matthaeum in lingua Hebraica*, Basel, 3rd. ed., 1582

Manfred Büttner, Dr. phil., Dr. rer. nat., Dr. theol., is Professor of the History of Geography and of Cultural Geography at the Ruhr-Universität, Bochum, West Germany, and Karl Heinz Burmeister, Dr. phil., Dr. jur., is Director of the Provincial Archives of Voralberg, Bregenz, Austria.

DATES	LIFE AND CAREER	ACTIVITIES, TRAVEL FIELDWORK	PUBLICATIONS	CONTEMPORARY EVENTS AND PUBLICATIONS
1488	Born 20 January in Niederingelheim, near Mainz			
1491				Copernicus was a student of Cracow University
1492				First globe by M. Beheim; Beginning of Columbus' expeditions
1499				Amerigo Vespucci's first voyage
1504				*Cosmographiae introducio,* St Dié
1505	Entered the Franciscan Order in Heidelberg			First history of Germany by J. Wimpheling
1507	Student of mathematics, geography and astronomy in Louvain			
1509	Pellikan's favourite student in Rufach			J. Calvin born in Noyon
1512	Ordination as a priest			Copernicus' *Commentarioulus;* G. Mercator born in Rupelmonde
1514	Lecturer at the Ordensstudium in Tübingen and student of Stöffler			
1517				Beginning of the Reformation in Germany
1518	Lecturer at the Ordensschule in Basel and at the Adam Petri publishing house			Melanchthon became a professor at Wittenberg University
1520	Sympathizing with Lutheran ideas		*Epitome hebraicae grammaticae,* one of many philological and grammatical works	
1522	Lecturer in Hebrew at the Ordensstudium in Heidelberg			Luther's translation of the New Testament
1524	Professor of Hebrew at Heidelberg University	Began to work on *Cosmography,* influenced by B. Rhenanus		Petrus Apianus, *Cosmographicus liber*
1526		Exploration of the course of the Rhine in the district of Heidelberg		
1528		Enlisted the co-operation of various specialists in describing Germany and Europe		New organizations of institutional and ecclesiastical affairs in Saxony

DATES	LIFE AND CAREER	ACTIVITIES, TRAVEL FIELDWORK	PUBLICATIONS	CONTEMPORARY EVENTS AND PUBLICATIONS
1529	Appointment as Professor of Hebrew at Basel University; left the Franciscan Order; married Adam Petrie's widow			
1530			*Germaniae descriptio; Hebraica Biblia Latina* (to 1533)	Sebastian Frank's *Weltbuch* the first popular cosmography
1534				
1535				Calvin visited Basel
1536			*Mappa Europae*	Calvin's *Institutio religiones christianae*
1537		First exploring expedition to Swabia		
1538			Translation of Tschudi's *Rhaetia;* Editions of Solinus and Mela	Calvin went to Strasbourg
1540			Latin edition of Ptolemy's *Geography*	
1541		Second expedition to Swabia		Calvin returned to Geneva
1542	Professor of Old Testament Theology (to 1544)			
1543		Third expedition to Swabia		Death of Copernicus
1544		Foundation of 'complete geography'	*Cosmography*	
1546	Rector of Basel University (to 1548)	Expedition to southeastern Swabia and Valais		
1550			Final publication of the *Cosmography*	
1552	Died of the plague, 26 May, in Basel			

Robert Swanton Platt

1891–1964

RICHARD S. THOMAN

Robert Swanton Platt was born in Columbus, Ohio, on
14 December 1891, and died in his home in Chicago,
Illinois, on 1 March 1964. His lifetime spanned the
time of initiation and establishment of geography as
a university discipline in the United States. In
cooperation with a comparatively small number of dedicated colleagues, Platt was and is responsible for
some fundamental contributions to the field as it
exists today.

1. EDUCATION, LIFE AND WORK
Platt was the eldest in a family of four children that
included two brothers and a sister. His father, a
lawyer of financial means in Columbus, encouraged him
to become a lawyer also. All of the family were
members of the Episcopal Church, and all appear to
have been quietly but sincerely devout. His extended
family had roots in New England, and he returned
occasionally to Kennebunk, Maine, and other places in
New England. After completing his elementary education in Columbus, he enrolled for high school at
the St George School in Newport, Rhode Island, and
later at the Hotchkiss School in Lakeville, Connecticut. New England was thus important in terms of
environment during his adolescent years, as well as
in terms of family and cultural heritage.

Throughout youth and adolescence, Platt was encouraged into a wide range of mental and physical
activity. The subject matter at the St George and
Hotchkiss schools emphasized traditional themes:
classical languages and literature, modern languages
and literature, mathematics, history, religions. At
St George, as an outstanding scholar in Latin, he won
as a prize a leather-bound set of Shakespeare's
complete works. His range of interests included
music and sports. In his early years, it was not
uncommon for his family to review outstanding works
of literature as an evening pastime, with one member
reading aloud. This practice remained with him, and
was continued by his own family. He often visited
museums, art galleries and other cultural centres.
He was an acute observer and recorder of events in his
lifetime, maintaining a concise but nearly continuous
diary from 1910 until his death.

In the autumn of 1910, shortly before his 19th
birthday, Platt entered Yale University. Yale's
curriculum was in many ways a continuation of the
traditional studies of his earlier years, plus such
new subjects as economics, physics, geology and
philosophy. On 27 January 1911 he heard an address
by Gifford Pinchot on 'Advance in conservation' and
noted the enthusiastic reaction of the audience.

Platt was also an enthusiastic traveller. On
19 June 1913 he set out with his father and immediately
younger brother on a trip across the United States
and Canada, through Chicago, Winnipeg and Banff to
Vancouver, where they sailed on the *Empress of
Russia* to the Far East. On 24 June they crossed
the 180th Meridian, and on the following day were
near the Aleutian Islands. Landing in Yokohama on
30 June, they visited several cities in Japan before
proceeding to China early in July. In China, they
went to Shanghai, Soochow, Nanking, Hankow, and
Changsha before proceeding to Mukden in Manchuria

and Seoul in Japanese-controlled Korea.   They sailed for home in late August, arriving in Columbus on 26 September.   Within days he returned to Yale for his final undergraduate year.

On 15 February 1914 Platt recorded that he had been appointed to teach the following year at 'Yale in China', located in Changsha.   He then was 22 years of age.   After graduating from Yale in 1914 with a major in philosophy, he sailed with his family for Norway. Having travelled as a unit in Scandinavia, the family parted, with Platt and his father continuing to St Petersburg in Russia.   They reached the Baltic seaport on the evening of 20 August, and Platt departed alone the following day on the Trans-Siberian Railroad for Siberia in Czarist Russia. (Hartshorne states that Platt was detained in Russia as a possible spy on this journey (Hartshorne, 1964)). On 9 September he reached Mukden again, thus -- as noted in his diary -- completing a trip around the world within one year and one month.   On 16 September he reached Changsha, and two days later was assigned his 'first recitations'.   These amounted to a total of 23 hours per week, and involved two subjects: the Bible and geography.   For the first time, Platt encountered geography as a formal subject.   Although his background and course work in the Bible had been ample, his only preparation for his geography responsibilities had been one course in geology at Yale.

He began to develop an interest in geography during this year, about which he later wrote:
Remembering the geological field trips in Connecticut, I took my geography class of Chinese boys on trips into the countryside of Hunan Province.   They were interested and stimulated, and my curiosity was aroused as I analyzed the striking similarities and striking contrasts between the Chinese landscapes and those of New England.   Moreover, both China and New England were different from other regions seen during the journey to the Far East via Europe and Siberia, and during the return to the United States via Japan and Hawaii. (Hartshorne 1964)
Returning to the United States after completing his year of instruction, Platt discussed his experiences in China and his interest in landscape comparisons with a travelling companion, who said: 'My brother-in-law, Wellington Jones, who is teaching at the University of Chicago, is interested in much the same sort of thing. *He* calls it all Geography' (Hartshorne, 1964).   Platt went to the University of Chicago and enrolled in the relatively new Department of Geography, which had been established there by the geologist Rollin D. Salisbury in 1903.   On 29 September 1915 he recorded that he 'took a room with Mrs Elizabeth Hoyt at 5704 Kenwood Avenue'.   Mrs Hoyt was the mother of Homer Hoyt, who was to become renowned for his various contributions to land economics and geography.   Platt and Hoyt exchanged ideas frequently and played tennis from time to time.   There does not appear to be much evidence that either influenced the other's thinking appreciably.   Mr Hoyt has explained in personal conversation that the time of their interaction was early in both of their professional careers.

Later, Platt recalled his reasons for choosing geography as a discipline and a career:
....In comparison with my philosophy major, geography offered the advantage of dealing with tangible and visible things forming a solid basis on which to build ideas, instead of beginning with my history minor, geography had the advantage of going more to the field for direct observation instead of going to the library to read about things no longer visible.   In comparison with geology, geography had the appeal of dealing with the world of people instead of only rocks and fossils. (Hartshorne, 1964)
He recorded attendance at lectures by Harlan H. Barrows on the influence of geography on history and on conservation of natural resources; by Walter Tower on political geography and meteorology; by Rollin D. Salisbury on geographical geology and the geology of continental evolution; by J. Paul Goode on economic geography; by Charles Colby on North America.   He noted with particular enthusiasm various field trips with Wellington Jones.   He mentioned Ellen Churchill Semple, who lectured at Chicago for portions of alternate years while he was a graduate student there.

His graduate work was interrupted in 1917 by a period of military duty, but early in 1919 he resumed his studies with the Department of Geography at Chicago.   In that year, prior to receiving his Ph.D. degree, he was appointed to the faculty of the Department, meeting his first classes in January 1920.   Those classes were elements of geography and North America.   Tower had left the Department by the time Platt joined, but Derwent Whittlesey had been added, together with Platt.   In December 1920 Platt received his Ph.D. degree in geography from the University of Chicago.   In 1921 he met Richard Hartshorne, who was to become a lifetime friend and professional colleague.

In December 1922 Platt married Harriet Shanks, who had been a student in some of his classes; and, from that day, the two worked essentially as a team in a wide range of professional and personal activities, at times with their two children, Robert ('Little Bob') and Nancy.   Their large house at 10820 South Drew Avenue became a base to which students, faculty and friends were invited to engage in lively and animated discussion.   The couple planned and carried out very many field excursions, some designed mainly for research purposes, and others for teaching.   Both types of excursion contributed substantially to Platt's growing reputation as a field analyst.   The Platts' interest in people extended well beyond scientific analysis to include subjective compassion and desire to assist others. During their marriage of almost 42 years, a total of 127 young persons came to live in their home.   Most had met the Platts in one of the couple's field trips into Latin America, Southeast Asia, Europe, or elsewhere.   While living at the Platt home, many were studying at the University of Chicago or other educational or training institutions in the Chicago area.   They came to be known as 'Plattachés'

-- a term chosen by the group itself to include only those who had lived one year or more as members of the 'family'. However, young people were not the sole recipients of assistance from the Platts. From 1929 to 1934, the early depression years, Platt served as Treasurer of the Association of American Geographers; in that capacity he made several financial contributions to the organization.

Fieldwork in geography clearly was the highest in Platt's list of priorities. Beginning in 1920, he conducted a field course almost every summer, at first in the upper Great Lakes Region, then along the international boundary of the United States and Canada, and finally as far northward as James Bay. Between 15 and 20 students usually were members of these field trips, which came to be required of all students receiving a higher degree from the Department. Each trip lasted for approximately one month. Specific objectives changed somewhat as Platt's own views changed. In the earlier work, efforts were made to gain direct insight into reciprocal associations involving people, space, and the social and physical settings. This was accomplished by meticulous mapping of very many such associations, at the micro-level of observation. Although he had attended lectures by Ellen Churchill Semple, Platt consistently reacted against what he conceived to be a strong theme of environmental determinism in her work. The agenda he placed before his summer field camp students therefore was, by design, an objective search for associations which could be observed and classified, whether these involved people, people and the physical setting, or the physical setting alone. However, Platt consistently regarded the physical setting in terms of its implications for human occupance, and not as an entity to be considered apart from such implications. In his later years, his interest turned to the human organization of space, considered in view of, but not determined by, past historical events and the physical setting. Those summer field trips were carried out under Platt's direction for more than thirty years.

Meanwhile, geography had been expanding as a discipline, especially in the Midwest, the East, and the West Coast. The Association of American Geographers had been founded by William Morris Davis and others in 1904, one year after the initiation of the Department of Geography at the University of Chicago. Commencing in 1923, a series of annual field conferences were held to facilitate the discovery and exchange of ideas on critical topics of that time.

These conferences, which were organized by interested geographers and were not part of offical AAG activities, continued for 17 years (James, 1972, 1977). Hartshorne states in private correspondence that the initial group, eleven in number, were called together by Wellington Jones and Carl Sauer, meeting nine times between 1923 and 1932. A group of younger colleagues also became interested, and after 1935 the two groups joined for five sessions. Platt and Whittlesey attended all 14 meetings.

The first five conferences dealt chiefly with procedures and objectives of field surveys in agricultural areas. The next seven involved experimentation with methods such as the field traverse, the role of microgeography, and the relationship between specific features and general classifications. The last five were concerned especially with broader objectives -- land classification, regional planning, ecological relationships.

Platt attended these conferences regularly. Some others included R.H. Brown, C.C. Colby, D.H. Davies, S.N. Dicken, S.D. Dodge, L. Durand, V.C. Finch, W.H. Haas, R.B. Hall, R. Hartshorne, P.E. James, W.D. Jones, H.M. Leppard, A.K. Lobeck, K.C. McMurry, A.E. Parkins, C.O. Sauer, G.T. Trewartha, J.R. Whitaker, and D.S. Whittlesey. Platt referred frequently to the conferences in the development of his own approach to fieldwork.

Most of Platt's personal fieldwork, again usually with the assistance and very close cooperation of his wife, Harriet, was conducted in Latin America. His choice of this area was partly the result of circumstances when he joined the Chicago faculty: no one there had a major interest in Latin America at the time he became a member, and it was believed that such an interest should be developed.

Chauncy Harris, in personal correspondence, has a further explanation:

I recall his reporting on a number of occasions that there was a curious reversal of regional interests by members of the faculty. When Wellington D. Jones was asked to join the faculty of the Department of Geography, after having spent three years with Bailey Willis in Argentina, Jones might naturally have become the Latin American specialist but Walter S. Tower had already pre-empted that regional specialty, so Jones, who had never been to Asia, decided to become the Asia specialist. Later on when Platt, who had been at Yale in China and travelled on the Trans-Siberian, was asked to join the faculty, he might by experience have specialized in Asia, but by then this area was already covered by Jones, so he took Latin America, since Walter S. Tower had resigned from the Department to become a Commercial Attaché for the United States Government after having been on leave during World War I for work in Washington.

Between 1922 and 1941 the Platts made seven trips to Latin America, each lasting from two to six months. These field trips involved micro-studies and some traverses which aggregately touched parts of nearly every major region in Latin America. They included a very wide range of both latitude and altitude, and focused chiefly on micro-studies of people, organized and functioning in units which ranged in size and scale from small trading posts in the Amazon to high mountain settlements in the Andes and commercial farming in Mexico. In all, 94 such study sites were utilized as a basis of generalization in his summary volume, *Latin America: Countrysides and United Regions*. After 1945 the couple made two more field excursions to Latin America, one by invitation from the Conselho Nacional de Geografia of Brazil to offer advice on selection of a site for the proposed Federal Capital (now Brasilia), and a second to study selected aspects.

of functional organization in Tierra del Fuego, at the southernmost tip of the continent, one of the few areas not visited in earlier work.

In the 1950s Platt's interest shifted regionally to north western Europe, particularly to field studies on the borders of the Netherlands, the German Federal Republic, and the Saarland, and the antecedents of those current boundary lines. In the early 1960s the Platts spent a year in Pakistan on Fulbright-assisted research, treating comparative occupance in tropical areas.

Platt remained on the faculty of the Department of Geography at the University of Chicago from 1919 until retirement in 1957, serving as Chairman from 1949 to 1957. He also was Vice-Chairman of the Division of Geology and Geography of the National Research Council from 1937 to 1939, Advisor to the Geographic Office of the Department of State in 1943, Chief of the Division of Maps, Library of Congress from 1944 to 1945. After retirement, he was Visiting Professor at the University of Frankfurt-am-Main, and at the University of the Saarland, and also lectured at several other universities in Europe. Platt's interest in Western Europe was stimulated by his participation in a faculty exchange project involving the University of Chicago and the University of Frankfurt-am-Main, in which Chauncy Harris had participated in 1950[1], when he was Chairman of the Department at Chicago, and Harris reciprocated later while Chairman of the Committee on the Chicago-Frankfurt exchange. Platt's interest was further stimulated during a Fulbright sojourn at Münster (especially with Professor Wilhelm Müller-Wille, who had been Visiting Professor at Chicago), and in his work at the Saarland. He spent periods as Visiting Professor at Rollins College, and at Indiana, Ohio State, and the Michigan State Universities (Harris, 1964). While in Pakistan, he lectured and conducted research at the University of Dacca. From 1961 until his death in 1964, he served as editor of the *Annals of the association of American geographers*. Among honours received by Platt in his lifetime were selection as Treasurer (1929-34), Vice-President (1943) and President (1945) of the Association of American Geographers; the Helen Culver Gold Medal of the Geographic Society of Chicago; honorary membership in the Gesellschaft für Erdkunde zu Berlin; and membership in the Società Geographica Italiana in Rome.

2. *SCIENTIFIC IDEAS AND GEOGRAPHICAL THOUGHT*

It is clear that field reconnaissance was important to Platt long before he chose geography as his discipline of professional interest, and that fieldwork provided the foundation for nearly all of his professional work. Of the 48 total contributions he made to the discipline, nearly all are based on field studies, each on a distinct field effort. He brought to that fieldwork an appreciation of nuances as well as observable circumstances involving the associations that relate human beings with each other and with other parts of the world. Yet, from the outset, he saw such associations as a congeries of unusual if not unique conditions rather than universal qualities easily generalized into large wholes. Certainly he appreciated the macro-perspective, but was essentially sceptical concerning what he considered to be facile and loose generalizations. He was especially wary of such generalizations when they appeared to be based on pre-judgements.

Although he entered geography at a time when environmental determinism was an important theme in colleague of Ellen Churchill Semple, one of the foremost proponents of that theme, his own work never emphasized that particular approach. His Ph.D. dissertation, 'Resources and economic interests of the Bermudas', completed in 1920 when environmental determinism was a significant aspect of geographical research, did not emphasize this theme. The opening sentence of the abstract to his dissertation sets a tone reflected in the entire work: 'The Bermudas are a unique group of islands: minute, bizarre, remarkable.' During his mature years, he became increasingly aware of the need to base generalizations on a foundation of micro-conditions and events, giving attention to the exceptions as well as to the parts which appeared to fit well into a large scheme or system. Describing this method, he would say, 'One must start with the large view, then treat with the small, then return to the large again' (Hartshorne, 1964). Stated in different words, his approach involved 'detailed work in a very small area combined with reconnaissance over a great area. The intensive field work deals with the intimate details, the extensive reconnaissance places the small area in the large, and spreads the detailed findings to build up regional generalizations' (Hartshorne, 1964).

In his fieldwork, Platt was becoming increasingly sensitive to human organization of space -- to areal units reflecting this organization at differing scales. A single unit of organization, in his view, involved a central point of decision making, with lines of interaction connecting that point with space with which that point had associations. Thus a farmstead is the focal point for a farm, with the lanes the lines of interaction; a city is the focal point for a trading area, with transportation and communication lines of interaction; a political capital is the focal point for a territory under its jurisdiction, again connected by interactional lines of transportation and communication; an international node of population with high economic demand is the focal point of world trade, with global trading lanes, involving both transportation and communication, providing interactional linkages.

This concept first appeared in Platt's work in 1928, in a short field study of Ellison Bay, Wisconsin, where a small monument has been erected to commemorate students from all parts of the world who had come to Ellison Bay to study with Platt. Recognition of the unit of spatial organization, which may or may not be aggregated into a nodal region, may well have been the single most important methodological contribution by Platt to the field of geography. Harris states:

> This Ellison Bay paper viewing communities and regions as 'points of focus, areas of organization, and lines and limits of movement, rather than static areas of uniformity' initiated among geographers a new way of looking at man's organization of space. These new concepts

continued to develop and gain momentum as they evolved into studies of areal functional organization. They contributed importantly to central-place theory and the study of economic regions. This paper was also the first intimation, as far as I know, of a clear discernment of the sharp distinction between regions of homogeneity and regions of organization (or focal regions). (Harris, 1964)

Hartshorne states:

This concept, which represents perhaps his most distinctive contribution to American Geography, appeared first in the Ellison Bay study of 1928 and was illustrated in varying form in each of the studies made with his field classes as published during the period 1928-35. (Hartshorne, 1964)

In private correspondence, Preston James writes:

Platt deserves the credit you have given him for developing the essentials of the focal region. But there can be no doubt that the idea of identifying 'the reach of the village institutions' and so defining a focal point and its perimeter came from the discussions at the field conferences.

In his Presidential Address, Edward J. Taaffe credited Platt with providing fundamental contributions to two schools of thought in United States geography -- area studies and spatial organization (Taaffe, 1974).

This new insight became fundamental to Platt's fieldwork, as is indicated in the study of Tierra del Fuego:

Tierra del Fuego is an area of dynamic occupance, reflected in a pattern of organization, composed of lines and points. The dots and other symbols on a map of rural and urban establishments make it not merely a dot map of static distribution, but a dynamic map of points of focus in a pattern of organization: ranches, towns, packing plants, sawmills. The lines are not boundaries of static homogeneity, but limits of activity (ranch boundaries and international boundaries), or lines of movement (such as roads). This pattern is dynamic, in being formed by people moving in coordination and occupying the area. (Platt, 1949)

3. INFLUENCE AND SPREAD OF IDEAS

Platt's life work emphasized original, first-hand research, largely expressed as field studies. His contributions to the literature of geography, amounting to 48 books, articles, reviews and commentaries, mainly involved such studies, many of which were read as papers before the Association of American Geographers and later published in geographical journals. On three occasions, however, he reviewed and summarized his past work and thinking. These were: (1) publication of his summary volume, *Latin America: countrysides and united regions* (1942); (2) delivery of his Presidential Address, 'Problems of our time', to the Association of American Geographers (1946); and (3) publication of a significant fieldwork summary, *Field study in American geography: the development of theory and method exemplified by selections* (1959).

The first was a major effort to place into perspec-

tive a number of years of effort in Latin America. A total of 94 studies in widely separated latitudes and altitudes in Latin America, many not previously published, were presented here. At the outset, Platt indicated that these were not to be considered merely as separate micro-studies, but instead as components of larger perspectives. Under a section in his introduction entitled 'The geographer's dilemma', he stated:

The complex character of great regions, complicated particularly by human appurtenances, which necessarily are included in geographic comprehension, cannot be immediately and totally perceived by one pair of human eyes...Regional complexities are those, not of an amorphous mass, but of an intricate pattern in which details fit together in significant combinations.

Having presented the field studies under groupings of countries, Platt then attempted in the two final chapters to generalize, in two directions: (1) to classify the field studies by habitats, and develop significant associations, and (2) to generalize the political structure of Latin America into a greater world order.

Organization of space, apparent in *Countrysides and united regions*, came to the forefront of Platt's attention in 'Problems of our time'. With one square mile -- Midway Airport, the only airport then in Chicago -- as his point of focus, Platt suggested a range of regional, national, international and global ties to the airport in the form of traffic. He considered transportation and communication as the vital links to different levels of political organization, and a wide range of problems of differing types and dimensions. Then, concentrating on an elementary school tucked into a corner of the square mile, he indicated the localized linkage patterns associated with that institution, and some of the problems to be found there. From these two perspectives he proceeded to discuss unity and disunity, and problems arising therefrom, at different levels of observation, and the role of geography in offering solutions.

Field study in American geography reflects a mind whose thoughts were not easily anticipated, whether as to timing or depth. Interpreting field study broadly 'to include any geographic work in which the author has familiarized himself directly with the area concerned', Platt considered eight types of field study which had been initiated in practice (by such pioneers as Lewis and Clark, who were commissioned by President Thomas Jefferson in 1804 to explore the new Louisiana Purchase in central and western United States), or developed as methodology in university research. These were: (1) the exploratory traverse, (2) the area survey, (3) explanatory physical geography, (4) explanatory human geography, (5) analytical economic geography, (6) geography of areal uniformity and diversity, (7) geography of areal organization, (8) geography of cultural origin and dispersal. He evaluated pioneer efforts in each of these types in Part I, then traced derivatives of each (usually more than one) in Part II. With characteristic modesty, he included only one of his own studies -- that of Ellison Bay -- in Part I (under the geography of areal organization), and only seven of his field studies in Part II. The book contains excellent insight into some of the main threads of method as applied to fieldwork, and anticipates much of the literature now being produced, especially by the spatial structural school.

Throughout his career Platt retained his philosophical perspective. His concern with both ends and means was deep and continuing. Never attracted to determinism, whether environmental or social, he came to react strongly against both, and expressed that reaction in two articles written in 1948: 'Environmentalism versus geography', and 'Determinism in geography'. These articles were significant milestones in the shift of geography away from the excesses which motivated his writing of them. Some have maintained that Platt, in taking so strong a position against determinism, set the stage for the other extreme of indeterminism. (See particularly V.A. Anuchin as quoted in James, 1972, p. 299.) Geography in the 1930-50 period did move into a condition of qualitative interpretation which some observers called 'mere description'; and the rise of the spatial structure school was in part a reaction against the lack of precision and of effective generalization prevalent in much geographic work of that period.

Platt's contribution to geography also is reflected today in the thinking of very many of his former students and other associates. Taaffe has shown that some scholars in the spatial structure group have credited Platt with providing a foundation, especially in his work on the dynamic region, upon which systems analysis has been constructed (Taaffe, 1974). However, it is noteworthy that, although Platt's work is so conceived within the spatial structure school, which has concentrated research efforts on nodes and linkages that are essentially urban, most of Platt's own work was on rural areas and very small settlements. He never seemed to enjoy working with complex urban situations.

Platt's influence on others was a mixture of intellectual impact and personal concern, and the two qualities never were easily separated. In this respect, Hartshorne has written:

In his relations with colleagues, students or other persons with whom he came to converse, his pleasure was to seek to learn from their experiences and views. Having discovered that there was almost always something of interest in any idea and in any person, he was genuinely stimulated by a continual variety of human contacts, and he was stimulating to almost everyone he met. (Hartshorne, 1964)

One former student and colleague has written:

My personal opinions concerning Platt are as follows: (1) he exerted very little influence on the thinking of anyone who (a) didn't know him well and (b) wasn't sufficiently intelligent to see the significance of his ideas; (2) his undergraduate major in philosophy made him sufficiently knowledgeable philosophically to avoid the various elementary blunders that many of his geographic-philosophic colleagues perpetrated, and (3) his highly unorthodox ideas -- witness microgeography -- and the logically sophisticated but verbally simple language in

which he presented his ideas caused many of his readers and auditors to be completely opaque to his ideas. They thought, because his language was not heavily turgid, that his ideas were superficial, hence failed to try hard enough to understand them; in fact, didn't even realize that they had failed to understand. (Wesley Calef, Professor of Geography, Illinois State University, Normal, Illinois)

Another former student and colleague stated:

As a person, Robert Platt was remarkable, it seems to me, for the unity of his life. He was much the same person in the classroom and in the office, on campus and in the field, at conventions and at home. His work seemed always essentially enjoyable to him; conversely, his vacations were normally professional field trips. The famous openness at his house was matched to a considerable extent by the openness of his academic pursuits to his family.

As a geographer, we should remember, he was of the generation in the United States that was called upon to re-found geography which, once 'the study of the relation of earth and life', had declined in credibility. In a time of 'solutions' which often were rather individualistic, Platt was distinguished for the consistency with which he held to the faith that the experience of travel and visitation, if directed toward regional understanding, was the best school for geographers. (William D. Pattison, Chairman, Department of Geography, University of Chicago)

There are some similarities and some differences in a third viewpoint from a former student:

I have a few miscellaneous recollections: (1) Platt's habit of relentlessly pursuing a few overall themes in all his courses, writings, and student advising, so that his works had an usual degree of cohesion and consistency; (2) Platt's trenchant commentaries written in the margins of student papers, giving very valuable lessons in very few words (for example, 'needs historical-cultural-functional frame to justify physical analysis of the area'); (3) Platt's lecture in which he maintained that geographers have five basic 'handles' to use in going about their work: (a) natural setting, (b) areal homogeneity, (c) culture origin and dispersal, (d) sequent occupance and (e) functional organization. No new 'handles' had been developed for some time. (Jesse H. Wheeler, Jr, Professor of Geography, University of Missouri, Columbia)

Possibly a key to the life and work of Platt may be found in the following quotation from a paper entitled 'My religious exposure', presented by him in 1954 to an interdisciplinary group of faculty at the University of Chicago:

We are trying to understand the universe from the viewpoint of our life on earth and in this attempt we use one or more of the methods devised for the purpose.

One method is that of empirical science, elaborated effectively in the physical and biological sciences, and copied with less success and more false hope in the social sciences. Another method is that of conceptual thought and

logic, elaborated in mathematics, philosophy
and law, and used with more doubtful benefit in
religion ...

   Other methods seem to be needed for reaching
effectively some areas of knowledge, particularly
where tangibles and intangibles come together,
where human will and living values impinge on the
known world, where mystery is as real as reality,
where integration is more needed than analysis,
and commensurability is hard to find.   Here is
a twilight zone not clearly exposed to the steady
light of systematic learning from either direction,
inductive or deductive, but illuminated
intermittently by flashes of comprehensive understanding.   At different points in the twilight
zone are psychology, history, geography and
religion.
   The author expresses appreciation for written
commentary on this work from Wesley Calef, Chauncy D.
Harris, Richard Hartshorne, Preston James, J. Trenton
Kostbade, William D. Pattison, William L. Thomas, and
Jesse H. Wheeler, Jr.   Mr James Delehanty, a graduate
student at the University of Chicago interested in
Platt's career, also has been very helpful.

Bibliography and Sources

1. REFERENCES ON ROBERT S. PLATT
Harris, Chauncy D., 'Robert Swanton Platt', *Geogr.
   Rev.*, vol. 54, (1964), 444-5
Hartshorne, Richard, *Perspective on the nature of geo-
   graphy*, Assoc. Am. Geogr., (1959), 201 p.
Hartshorne, Richard, 'Robert S. Platt', *Ann. Assoc.
   Am. Geogr.*, vol. 54, (1964), 630-7
Hartshorne, Richard, *The nature of geography*, Assoc.
   Am. Geogr., (1939), 482 p.
James, Preston E., *All possible worlds: a history of
   geographical ideas*, Indianapolis, 1972, 622 p.
James, Preston E., and Cotton Mather, 'The role of
   periodic field conferences in the development of
   geographical ideas in the United States', *Geogr.
   Rev.*, vol. 67, (1977), 446-61
Taaffe, Edward J., 'The spatial view in context',
   Ann. Assoc. Am. Geogr., vol. 64, (1974), 1-16

2. SELECTED REFERENCES BY ROBERT S. PLATT
1920 'Resources and economic interests of the Bermudas'
   (Unpubl. dissertation; abstract in *University of
   Chicago Abstracts of Theses, Science Series*,
   vol. v, 325-30)
1932 'A cross section of Central America in Costa
   Rica', *J. Geogr.*, vol. 22, 95-100
1926 'Central American railways and the pan-American
   route', *Ann. Assoc. Am. Geogr.*, vol. 16, 12-21
1927 'A classification of manufactures, exemplified by
   Porto Rican industries', *Ann. Assoc. Am. Geogr.*,

   vol. 17, 79-91
1928 'A detail of regional geography: Ellison Bay
   Community as an industrial organism', *Ann. Assoc.
   Am. Geogr.*, vol. 18, 81-126
1930 'Pattern of land occupancy in the Mexican Laguna
   District', *Trans. Illinois State Acad. Sci.*,
   vol. 22, 533-41
1931 'An urban field study: Marquette, Michigan',
   Ann. Assoc. Am. Geogr., vol. 21, 52-73
1933 'Magdalena Atlipac: a study in terrene occupancy
   in Mexico', *Bull. Geogr. Soc. Chicago*, vol. 9,
   45-75
1935 'Field approach to regions', *Ann. Assoc. Am.
   Geogr.*, vol. 25, 153-172
1938 'Items in the regional geography of Panama,
   with some comments on contemporary geographic
   method', *Ann. Assoc. Am. Geogr.*, vol. 28, 13-36
1939 'Reconnaissance in British Guiana, with comments
   on microgeography', *Ann. Assoc. Am. Geogr.*,
   vol. 29, 105-26
1942 *Latin America: countrysides and united regions*,
   New York, 564 p.
1946 'Problems of our time', *Ann. Assoc. Am. Geogr.*,
   vol. 36, 1-43
1948 'Environmentalism versus geography', *Am. J. Sociol.*,
   vol. 38, 53, 351-8
   'Determinism in geography', *Ann. Assoc. Am. Geogr.*,
   vol. 38, 126-32
1949 'Reconnaissance in dynamic regional geography:
   Tierra del Fuego', *Rev. Geogr. Inst. Pan-Am.
   Geogr. Hist.*, vols. 5-8, 3-22
1952 'The rise of cultural geography in America',
   Proc. 17th Int. Geogr. Congr., 485-90
1957 'A review of regional geography', *Ann. Assoc. Am.
   Geogr.*, vol. 27, 187-90
1959 *Field study in American geography*, Univ. Chicago
   Dept. Geogr. Res. Pap., no. 61, 405 p.

3. UNPUBLISHED SOURCES ON ROBERT S. PLATT
Mikesell, Marvin W., 'Robert S. Platt: an
   appreciation' (Memorial service for Platt at
   Rockefeller Chapel, 17 March 1964; also includes
   remarks by three 'Plattachés'; in Platt collection
   at Univ. of Chicago)
Pattison, William D., 'An exercise in the sociology
   of knowledge' (Paper by then graduate student at
   Univ. of Chicago on Platt's life and contributions,
   written in 1952; in Platt collection at Univ. of
   Chicago)

*Richard S. Thoman is Professor of Geography at the
California State University, Hayward, California,
U.S.A.*

CHRONOLOGICAL TABLE: ROBERT SWANTON PLATT

DATES	LIFE AND CAREER	ACTIVITIES, TRAVEL FIELDWORK	PUBLICATIONS	CONTEMPORARY EVENTS AND PUBLICATIONS
1891	Born 14 December at Columbus, Ohio			
1903				Establishment at the University of Chicago of the first Department of Geography in the United States
1904				Organization of the Association of American Geographers
1910	Entered Yale University			
1913	Awarded membership of Phi Beta Kappa fraternity	Travelled with father and brother to western United States, Canada, Japan and China		
1914	Graduated from Yale with major in philosophy; accepted position as instructor in geography and the Bible at 'Yale in China', located in Changsha	Travelled again to China for academic year 1914-15		
1915	Enrolled in Geography Department at University of Chicago for graduate study			
1917	Military duty in First World War (to 1919)	Various locations in the United States		Participation by United States in First World War
1920	Appointed Instructor of Geography at University of Chicago; Received Ph.D. degree from University of Chicago Began conducting field courses at University of Chicago that continued throughout his career		'Resources and economic interests of the Bermudas' (dissertation)	
1922	Married Harriet Shanks			
1922	Began to invite 'Plattachés' to his home in Chicago	With Harriet, made seven field trips to Latin America, each lasting from two to six months (to 1941)		
1923			'A cross section of Central America in Costa Rica'	Initiation of annual field sessions involving geographer from many U.S. universities

DATES	LIFE AND CAREER	ACTIVITIES, TRAVEL, FIELDWORK	PUBLICATIONS	CONTEMPORARY EVENTS AND PUBLICATIONS
1926			'Central American railways and the pan-American route'	
1927			'A classification of manufactures, exemplified by Porto Rican industries'	
1928			'A detail of regional geography: Ellison Bay community as an industrial organism'	
1929	Treasurer, Association of American Geographers (to 1934)			
1930			'Pattern of land occupancy in the Mexican Laguna District'	
1931			'An urban field study: Marquette, Michigan'	
1933			'Magdalena Atlipac: a study in terrene occupancy in Mexico'	
1935			'Field approach to regions'	
1937	Vice-Chairman, Division of Geology and Geography, National Research Council (to 1939)			
1938			'Items in the regional geography of Panama, with some comments on contemporary method'	
1939			'Reconnaissance in British Guiana, with comments on micro-geography'	
1942			*Latin America: countrysides and united regions*	Participation by United States in Second World War
1943	Vice-President, Association of American Geographers; Advisor to Geographic Office, U.S. Department of State			
1944	Chief, Division of Maps, U.S. Library of Congress (to 1945)			

DATES	LIFE AND CAREER	ACTIVITIES, TRAVEL, FIELDWORK	PUBLICATIONS	CONTEMPORARY EVENTS AND PUBLICATIONS
1945	President, Association of American Geographers	Two field trips to Latin America 9to 1947)		
1946			'Problems of our time	
1948			'Environmentalism versus geography'; 'Determinism in geography'	Shift of geography away from environmentalism and determinism aided by these two articles
1949	Chairman, Department of Geography, University of Chicago (to 1957)		'Reconnaissance in dynamic regional geography: Tierra del Fuego'	
1951			'Introductory field study'	
1952	Exchange Professor, University of Frankfurt-am-Main (to 1953)	Travelled to northwestern Europe	'The rise of cultural geography in America'	
1957	Visiting Professor, Universities of Frankfurt-am-Main, Münster, and the Saarland (to 1958); Retired as Emeritus Professor	Travelled to northwestern Europe	'A review of regional geography'	Early manifestations of the spatial structure school of geographic though based partly on Platt's earlier work
1959			*Field study in American geography*	
1960	Visiting Professor, Rolins College, Ohio State University (to 1961)			
1961	Visiting Professor, University of Dacca, East Pakistan (now Bangladesh) (to 1962); Editor *Annals of the Association of American Geographers* (to 1964)	Travelled to East Pakistan		
1962	Visiting Professor, Universities of Indiana and Michigan State (to 1963)			
1964	Died 1 March in Chicago, Illinois			

John Wesley Powell
1834–1902

PRESTON E. JAMES

By permission of the Smithsonian Institution

John Wesley Powell played a role of heroic proportions in the age of American pioneering. No school of geography having yet been established in America, Powell had no guidance or previous training on identifying geographical problems or setting about finding new answers. But he had an insatiable curiosity about the blank places on the maps, and about the ways of living of the inhabitants of those areas. In 1869 he led a small party down the Colorado River in boats, running rapids before the techniques of 'white-water boating' had been developed. Powell then organized and led survey parties into the arid lands of the West. He became a crusader for change in the settlement process, and for reforms that would protect the natural resources of the arid lands from destructive exploitation for private gain. In the 1880s he was at the peak of his power as director of the United States Geological Survey, authorized, briefly, to carry out and enforce a programme of land classification and a survey of irrigation potential. The story of his defeat by political forces opposed to reform, and to the dissemination of the kind of information he had collected, is a tragic one.

1. EDUCATION, LIFE AND WORK

John Wesley Powell was born on 29 March 1834 in Mt Morris, New York, a village on the Genesee River, some 34 miles south-southwest of Rochester. His father was an itinerant preacher in the Methodist Episcopal Church who kept moving his family westward beyond the area of established churches. In 1838 the family moved to the frontier village of Jackson, 25 miles southeast of Chillicothe in southern Ohio.

In these small pioneer communities there was a lack not only of churches, but also of schools. Young Wes had to pick up an education where he could find it. He was fortunate, for near his home in Jackson lived an amateur scientist named George Crookham, who agreed to tutor him in a variety of subjects. Crookham would suggest a book, and after the boy had read it the two would discuss the author's ideas. Before the age of ten, he had become an avid reader. But Crookham was also an out-door man who took him on long hikes to observe rocks and minerals, plants and animals, surface features and rivers, and also the way the inhabitants of this pioneer area were using, or misusing, the earth's resources. The two examined some nearby Indian mounds, looking for Indian artifacts, and also went out with the State Geologist to identify and map mineral resources. At ten years of age the boy was already familiar with the way professionals would go about exploring an area and mapping its resources.

In 1846 the family moved still farther west, to a farm in southern Wisconsin. Since the father was often absent on his rounds as a preacher Wes, the eldest son, had the responsibility of running the farm. After the harvest, he would drive a wagon load of wheat some five days to a port on Lake Michigan, where he would sell the wheat and buy needed supplies. This he was doing at the age of 14. When the family moved to Wheaton, Illinois, in 1852, Powell had a chance to attend Wheaton College, but he soon found

that the college work was too elementary. He began
to earn his own living as a schoolteacher, working at
nights to keep ahead of his pupils in subjects such as
grammar and mathematics. He used his summer vacations
to take long trips by rowing-boat on the Illinois River
and up and down the Mississippi, where he made collections of shells and plants which he could later sell.

As the Civil War approached, Powell began to
study map-making, military engineering and tactics.
In 1861 he joined the Illinois Volunteer Infantry as
a private, but within six weeks he was commissioned a
2nd lieutenant, and in four months he was promoted to
a captain. It was in this period that he married his
cousin, Emma Dean, who was ready to leave the security
of her home to share in the excitement of her husband's
exploring expeditions. But first came a period of
military service. Captain Powell commanded a battery
in the Battle of Shiloh in April 1862. In this
engagement his right arm was struck by a bullet and
had to be amputated below the elbow. After recovering
from surgery he rejoined his battery and took part in
the siege of Vicksburg under General Grant. Promoted
to major in 1864, he was Chief of Artillery in the
Battle of Nashville. In January 1865 he was dis-
charged from the army and returned to civilian life.

For a short time he held the post of Professor
of Geology, first at Illinois Wesleyan University,
then at Illinois State Normal University. But the
academic life was not for him, and he planned and
organized an expedition of volunteers to climb some
of the mountains in Colorado. Seeking financial aid
for his exploring parties he went to Washington.
His friend, General Grant, then Secretary of War,
arranged to have Powell's expeditions furnished with
rations from army posts, and the Smithsonian Institution loaned several surveying instruments. After
two field seasons among the Colorado mountains,
during which he climbed Pike's Peak and Long's Peak,
Powell began to plan an expedition to follow the
Colorado River through the midst of the largest blank
area of the map of the United States. On 24 May 1869,
the expedition launched its four boats in the Green
River at Green River Junction in Wyoming.
They followed the river through the Uinta Mountains,
and then on to the junction of the Green and the
Colorado before passing through the Grand Canyon and
reaching the Mormon settlement on the Rio Virgen on
29 August. By the time that the Major, as he was
usually called, had returned to Salt Lake City and had
taken a train back east, the newspapers had already
made him famous.

In July 1870 the 'Geographical and Topographical
Survey of the Colorado River of the West and its
Tributaries' was officially established in the
Department of the Interior, supported by annual
appropriations from Congress. Eventually, Powell's
survey was called 'The United States Geographical and
Geological Survey of the Rocky Mountain Region'.
He recognized that here was a vast, little-known
country about to be occupied by miners, lumbermen,
cattlemen, and farmers who had little idea of the
physical character of the land. It was of pressing
importance, Powell reported to various congressional
committees, that an accurate inventory of the
resources of the western territories be provided,
including a classification of the land in terms of

its potential use.

In 1879, acting on a recommendation of the
National Academy of Sciences, four separate surveys of
the West, including the one headed by Powell, were
combined in the United States Geological Survey, which
was directed to prepare a topographic map covering the
whole country, to complete the mapping of geological
formations, and to undertake a classification of the
public domain in terms of potential uses. At the
same time the Congress established the Bureau of
Ethnology in the Smithsonian Institution. Clarence
King (1842-1901) was named the first Director of the
Geological Survey, and Powell became the Director of
the Bureau of Ethnology. But in 1881 King resigned,
and for the next 13 years Powell directed both agencies. When he retired from the Geological Survey in
1894, he retained his position with the Bureau of
Ethnology.

In 1896, after recovering from another operation
on his arm, he retired to a cottage in Haven, Maine.
He and his wife returned to Washington each winter
to enjoy the intellectual life of the capital in
which he played a prominent part. When he died
at Haven on 23 September 1902, the estate he left
to his wife was very small. There were, however,
enough members of Congress who were Powell's friends
to pass an appropriation of $20 a month (later in-
creased to $50) as a pension for Emma Powell who
outlived him by 22 years.

2. SCIENTIFIC IDEAS AND GEOGRAPHICAL THOUGHT

a. Studies in geography and geomorphology

Powell's field studies between 1869 and 1879 were
focused on the Colorado Plateau and the Uinta
Mountains. As a result, not only was the last
large blank area on the map completed, but also a
new understanding of the processes of erosion by
running water and of the resulting landforms was
reached. Perhaps Powell's chief contribution to
the understanding of the processes of river erosion
was the concept of base level. Rivers, he recognized, could not cut below the level of the
standing water into which they emptied. The steeper
the slopes, the greater the speed of erosion; but,
given a sufficient length of time, continued erosion
would produce a plain of slight relief. Later
W.M. Davis called this a 'peneplain'; but the idea
of the peneplain was largely anticipated by Powell.

Powell also presented a new genetic classification of mountains and rivers. He described and
named consequent, antecedent, and superimposed
rivers. Grove Karl Gilbert, who worked with Powell
in this field, was able to expand Powell's ideas
about the sculpture of landforms, and to be far
more precise in measuring the transportation of debris
by running water. Gilbert is credited with developing the concept of an equilibrium between slope,
volume of water, and load of debris, which he called
'grade' but he was generous in crediting many of
his ideas to the informal discussions he had enjoyed with Powell, both in the field and later in
Washington.

b. *Studies in ethnology*

Ever since his examination of the Indian mounds in Ohio, Powell had been fascinated by Indian culture, and since there was no established field of ethnology to guide him, he had to blaze new trails in the preservation of Indian languages, customs, and mythology. During his field studies in 1868 and 1869, Powell compiled a list of seven hundred Ute words, and during the following years he continued to enlarge his knowledge of the different Indian languages. He was one of the few explorers of the West who could talk to the Indians in their own languages, but he was not interested in such studies for purely scholarly purposes. He wanted to aid the Indians to survive the difficult transition from tribal life to the white-man's way of living.

c. *Protector of the public domain*

No part of Powell's life is more spectacular than his heroic efforts to preserve the public domain from pillage for private gain. The programme he offered for land management in the arid lands was based on sound knowledge of the nature of the land and the needs of the people.

Powell attacked the whole idea of the rectangular land office survey. When farm or ranch properties are laid out in squares, as they are in the humid plains of Illinois, Indiana, and Ohio, most of the properties in arid lands will have no access to water for irrigation. The water, he pointed out, is not arranged in rectangles, but along the drainage lines. An entirely new method of land survey was needed to provide water for a maximum number of owners. He even urged that the political divisions in the arid lands should be so drawn that watersheds came under one political authority.

Although Powell had probably never read Humboldt, he was one of the first in the Western territories to understand the relation of a cover of woodland on mountain slopes to the flow of water to bordering lowlands. In 1878 Powell's *Report on the lands of the arid region of the United States* was published by the Department of the Interior. (A second printing, with additional chapters, appeared in 1879. The book was reproduced with an introduction by Wallace Stegner in 1962 by Harvard University Press.) Stegner calls it 'one of the most remarkable books ever written by an American'. It describes the character of the arid lands, and offers a specific programme for avoiding the destruction of the land by improper use, even going so far as to suggest the form of the bills that might be passed by Congress for the more effective use of these arid lands.

All these proposals met with vigorous opposition. The lumbermen wanted to cut trees without the extra cost of replanting. The cattlemen wanted to replace both forests and woodlands with grasslands suitable for summer grazing, and they demanded free access to the public domain for this purpose. The politicians who represented these various groups did not want any agency of the government publishing reports that would decrease the flow of new settlers; and certainly they did not relish a new survey of county boundaries, which would separate them from their constituencies. On the other hand, Powell found support for his proposals among some members of the Congress and in the National Academy of Sciences. He sought to broaden the base of his support by writing articles in the public press, and by public lectures.

Powell also faced another problem. It had become a popular myth in the West that as the land is ploughed rainfall will increase. Ferdinand V. Hayden, the leader of one of the other independent surveys of the West, had expressed support for this idea. Since this was a period of more than average rainfall, it seemed quite clear that ploughing the land would, in fact, cause an increase of moisture. Powell warned people of the inevitable return of a cycle of dry years, but this was a difficult point to make in a convincing manner among settlers who wanted to believe that the increase of rainfall was permanent.

In the late 1880s the drought that Powell had predicted set in. Many of the lands that had been confidently settled had to be abandoned. Suddenly there was pressure on Congress to authorize funds for a survey of irrigation possibilities. In October 1888 a generous appropriation was passed and Powell had what he had been seeking: the public domain was temporarily closed, and could only be opened by a Presidential order: and Powell was authorized to organize and carry out a survey of water resources. In August 1889 Powell was even invited to accompany the Senate Select Committee on Irrigation in a tour of the Western territories. But when the senators heard Powell urge that 'the sources of water should be kept in the hands of the people' they were shocked and appalled at what they had started. In October 1889 a crusade in the Congress was opened to destroy the irrigation survey and its director. The appropriation for the irrigation survey was cut down in 1891 and in 1892 the whole annual appropriation for the Geological Survey was cut nearly in half. Although some of the Geological Survey funds were restored the next year, Powell was no longer in a position of power. After he resigned in 1894 he devoted himself to studies of ethnology and to the writing of philosophy.

3. *INFLUENCE AND SPREAD OF IDEAS*

In the 1880s Powell was probably the most influential scientist in Washington. He was involved in the promotion of many different kinds of scientific studies by agencies of the Federal government, and he became the centre of intellectual circles in the nation's capital.

One of the changes Powell proposed was the concentration of the government scientific work in one new Department of Science. The National Academy of Sciences supported the plan; but there were also many influential scholars who felt strongly that the government should leave the promotion of research to the universities and other independent research agencies. After years of debate the proposal to establish a Department of Science was defeated. It is interesting that in 1950 the National Science Foundation was set up by the government to accomplish many of the purposes Powell had advocated.

It cannot be said that Powell was a great scientist. Scholars in the many fields to which he contributed agree that he was too quick to jump to conclusions while still lacking solid supporting evidence, and that he was not patient enough to undertake the detailed observations that might support his conclusions. But Powell's ability to foresee events far beyond the vision of most of his contemporaries was extraordinary. He predicted the growth of vast chemical industries based on petroleum at a time when this resource was only useful for lubrication and illumination. He foresaw that the great rivers of the West would some day be harnessed to provide electricity, and he even selected the site where a dam could control the waters of the Colorado River. The site he selected was only a few miles south of where the Hoover Dam has since been built.

The conservation movement in the United States was based on ideas developed by Powell. That this movement today is concerned broadly with the management of natural resources for the public good can be credited to Powell, either directly or through the men who worked with him and absorbed his point of view. In the summer of 1902, when Powell had already suffered a heart attack, President Theodore Roosevelt signed into law an act of Congress creating the Bureau of Reclamation, for which Powell had worked so hard. The Forest Service, the Bureau of Mines, and even the Soil Conservation Service are in this group. And today there is a great popular movement to protect the environment from destruction by private interests. That is what Wesley Powell had been fighting for almost a century earlier.

The extent of Powell's influence is revealed in the many honours he received. His lack of any advanced degrees from a university was remedied in 1877, when Illinois Wesleyan University awarded him both the M.A. and Ph.D. degrees. In 1881 Columbia University awarded him the LL.D. degree; and in 1886 Powell was included among 42 distinguished scientists and men of letters who received honorary degrees on the occasion of Harvard's 250th anniversary. In the same year the University of Heidelberg awarded the Major an honorary Ph.D. In 1891 the Fifth International Geological Congress was held in Washington, and after regular sessions were concluded, Powell and Gilbert conducted a western excursion to show the European delegates the landscapes of the arid West, including a view of the Grand Canyon. Later that same year the *Académie des Sciences* in Paris presented the Major with the Cuvier Prize for the 'Collective works of the Geological Survey'. Illinois College gave him the LL.D.; and he was elected a member of the Société d'Anthropologie of Paris, and of the Gesellschaft für Anthropologie, Ethnologie und Urgeschichte in Berlin.

Geography as a professional field was only beginning to emerge in America during the last decade of Powell's life. The first opportunity for advanced study of the physical aspects of geography was offered in the late 1880s at Harvard, where Nathaniel Southgate Shaler, one of Powell's strong supporters, was head of the Department of Geology. Powell's influence on the development of the field of geography during the present century was transmitted indirectly through William Morris Davis,

for Powell's insight into the significance of base level was developed by Davis into the concept of the cycle of erosion. Also Powell's influence was felt in the continuing concern with land classification studies basic to the management of natural resources, described by Charles C. Colby as 'one of the most persistent interests in American geography'.

Bibliography and Sources

1. REFERENCES ON J.W. POWELL

Gilbert, G.K., 'John Wesley Powell,' *Science* (new ser) vol 16, (1902), 561-7

Merrill, G.P., 'John Wesley Powell', *Am. Geol.*, vol 31 (1903), 327-33

Wolcott, C.D., 'John Wesley Powell', *U.S. Geol. Surv. Ann. Rep.*, (1903), 271-87

Gilbert, G.K. (ed), 'John Wesley Powell : a memorial to an American explorer and scholar', reprinted from *The Open Court* vol 16 and vol 17, Chicago, 1903

Hobbs, W.H., 'John Wesley Powell 1834-1902', *Sci. Mon.*, vol 39, (1934), 519-29

Darrah, W.C., *Powell of the Colorado*, Princeton, (1951), 426p.

Stegner, W., *Beyond the hundredth meridian: John Wesley Powell and the second opening of the West* (with an introduction by Bernhard De Voto), Boston, (1954), 438p.

2. PRINCIPAL WORKS OF J.W. POWELL

a. Physical geography of the arid region

1867 'Exploration of the valley of the South Platte, Colorado and ascent of Pike's Peak, scientific expedition to the Rocky Mountains', *Illinois State Board Educ. Proc.*, 9-13

1873 'Some remarks on the geological structure of a district of country lying to the north of the Grand Canyon of the Colorado', *Am. J. Sci. Arts* (ser 3), vol 5, 456-65

1875 *Explorations of the Colorado River of the West and its tributaries*, Government Printing Office, Washington D.C., 291p.

1876 *Report on the geology of the eastern portion of the Uinta Mountains and a region of country adjacent thereto*, Government Printing Office, Washington, D.C., 218p.

1878 *Report of the lands of the arid region of the United States*, preliminary edition, Government Printing Office, Washington D.C. 2nd edition 1879 with a more detailed account of the lands of Utah (includes chapters by A.H. Thompson and W. Drummond Jr.), Government Printing Office, Washington D.C., 195p. Reprinted with an introduction by Wallace Stenger, University of Chicago Press, (1962)

1891 'Hydrography, Engineering, the Arid Lands and Irrigation Literature', *U.S. Geol. Surv., 11th Ann. Rep., Part II*, 1–289, 345–88

b. Physical geography and geology

1876 'Types of orographic structure', *Am.J. Sci.Arts,* (ser 3), vol 12, 414–28

1884 'On the fundamental theory of dynamic geology', *Science,* vol 3, 511–13

1886 'The causes of earthquakes', *Forum,* vol 2, 370–91

1888 'Methods of geologic cartography in use by the United States Geological Survey', *Int. Geol. Congr.,* Berlin, 221–40

1893 'The geologic map of the United States', *Trans. Am. Inst. Min. Eng.,* vol 21, 877–87

1893 'General work in taxonomy', *U.S. Geol. Surv. 14th. Ann. Rep. Part I,* 65–112

1895 'Physiographic Processes', *Natl. Geogr. Monogr.,* vol 1, 1–32: 'Physiographic features', *op. cit.,* 33–64: 'Physiographic regions of the United States', *ibid.,* 65–100

1898 'An hypothesis to account for the movement in the crust of the earth', *J. Geol.,* vol 6, 1–9

c. North American Indians

1877 *Introduction to the study of Indian languages,* Government Printing Office, Washington D.C., 104p, 2nd edition 1880, 228p.

1878 'The philosophy of the North American Indians', *J. Am. Geogr. Soc.,* vol 8, 251–68

1881 'Sketch of the mythology of the North American Indians', *Smithson. Inst. Bur. Ethnol. 1st Ann. Rep.,* 17–56: 'Wyandot Government – a short study in tribal society', *op. cit.,* 57–69

1891 'Indian linguistic families of America north of Mexico', *Smithson. Inst. Bur. Ethnol. 7th Ann. Rep.,* 1–142

1892 'The North American Indians', in Shaler, N.S. (ed.), *The United States of America: a study of the American commonwealth,* vol 1, New York, 190–272

d. Anthropology

1881 'On the evolution of language ...' *Smithson. Inst. Bureau Ethnol. 1st Ann. Rep.,* on limitations to the use of some anthropological data', *op. cit.,* 71–86·

1883 'Human evolution', *Trans. Anthropol. Soc. Washington,* vol 2, 176–208

1884 'The three methods of evolution', *Bull. Philos. Soc. Washington,* vol 6, 27–52

1885 'From savagery to barbarism', *Trans. Anthropol. Soc. Washington,* vol 3, 173–96

1888 'From barbarism to civilization', *Am. Anthropol.,* vol 1, 97–123: 'Competition as a factor in human evolution', *op. cit.,* 297–323

1890 'Evolution of music from dance to symphony', *Proc. Am. Assoc. Advance. Sci.,* 1–21 (presiden-tial address at the 38th Annual Meeting)

1896 'On primitive institutions', *Am. Bar. Ass. Rep. 19th Am. Meet.,* 573–93

'Relation of primitive people to environment, illustrated by American examples', *Smithson. Rep.* (1895), 625–37

1900 'The lessons of folklore', *Am. Anthropol.* (new ser.), vol 2, 1–36

e. Philosophical

1880 'Mythological philosophy', *Proc. Am. Ass. Advance. Sci. 28th Annual Meeting,* 251–78

1882 'Outlines of sociology', *Trans. Anthropol. Soc. Washington,* vol 1, 106–29

'Darwin's contributions to philosophy', *Proc. Biol. Soc. Washington,* vol 1, 60–70

1896 'Seven venerable ghosts', *Am. Anthropol.,* vol 9, 67–91

1898 *Truth and error, or the science of intellection,* London, 428p

1899 'Esthetology, or the science of activities designed to give pleasure', *Am. Anthropol.,* (new ser.), vol 1, 1–40: 'Technology, or the science of industries', *op. cit.,* 319–49: 'Sociology, or the science of institutions', *ibid.,* 475–509, 695–745

1900 'Philology, or the science of activities designed for expression', *Am. Anthropol.,* (new ser), vol 2, 603–37

1901 'Sophiology, or the science of activities designed to give instruction', *Am. Anthropol.,* (new ser.), vol 3, 51–79: 'The categories', *op. cit.,* 403–30

Preston E. James is Maxwell Professor Emeritus of Geography, Syracuse University.

DATES	LIFE AND CAREER	ACTIVITIES, TRAVEL, FIELDWORK	PUBLICATIONS	CONTEMPORARY EVENTS AND PUBLICATIONS
1834	Born 29 March at Mt Morris, N.Y.			
1838	Moved to Jackson, Ohio	Field trips in Ohio with George Crookham		
1846	Moved to Wisconsin			
1852	Moved to Illinois; Taught in a country school			
1855–6		Travelled up the Mississippi to St Paul and to Michigan, then by the Mississippi to New Orleans		
1858	Curator of Conchology, Illinois Natural History Society			
1860	Principal of Hennepin Public Schools	Lecture tour in Tennessee, Kentucky, Mississippi		
1861	Married Emma Dean; Became a captain in the Artillery			Civil War
1862	Lost arm at battle of Shiloh			
1863				Siege of Vicksburg
1864	Became a major in the Artillery			Battle of Nashville
1865	Discharged from the army; Began teaching geology			
1866	Professor of Geology in Illinois	Exploration began including exploration to Colorado Pike's Peak; continued to 1869		
1867	Curator of the Natural History Museum		'Report of the scientific expedition on the Rocky Mountains', to Illinois Board of Education	
1868	Professor at Illinois Wesleyan University (to 1869)	Second expedition, to Long's Peak		
1870	Joined the Geographical and Geological Survey, Rocky Mountain region	Worked until 1878 for Field Surveys of Land Resources and Indians; Grand Canyon trip: studied the last cliff dwellers		
1871		Second Grand Canyon trip		
1873	Moved to Washington	Survey of plateaux of Utah, Arizona and Nevada		

DATES	LIFE AND CAREER	ACTIVITIES, TRAVEL, FIELDWORK	PUBLICATIONS	CONTEMPORARY EVENTS AND PUBLICATIONS
1875			*Explorations of the Colorado River*	
1876			*Report on the geology of the eastern portion of the Uinta Mountains*	
1877	Illinois Wesleyan University awarded him the M.A. and Ph.D degrees	Lectured on 'The public domain' to the National Academy of Science	*Introduction to the study of Indian languages*	
1878			'The philosophy of the North American Indians'; *Report on the lands of the arid region*	Cosmos Club founded
1879	Helped to found U.S. Geological Survey, with Clarence King as Director; Joined Bureau of Ethnology in the Smithsonian Institution and became its Director			
1881	Became Director of U.S. Geological Survey on the resignation of C. King; LL.D. degree of Columbia University		'Sketch of the mythology of the North American Indians'	
1884			'The three methods of evolution'	
1886	Received honorary degrees at Harvard and at Heidelberg			
1888		Approval of his proposed Irrigation Survey		National Geographic Society and Geological Society of America founded
1889	President, American Association for the Advancement of Science	Addressed the North Dakota Constitutional Convention; toured West with Senate Select Committee on Irrigation		
1890			'Evolution of music' (Presidential address to American Association for the Advancement of Science)	
1891	Cuvier prize given by Académie des Sciences, Paris; LL.D. degree, Illinois College	With G.K. Gilbert directed a tour of the West for visiting European geologists; state funds for irrigation reduced	'Indian linguistic families'	5th International Geological Congress, Washington
1892		Geological Survey funds reduced		
1893			'The geologic map of the United States'	

DATES	LIFE AND CAREER	ACTIVITIES, TRAVEL, FIELDWORK	PUBLICATIONS	CONTEMPORARY EVENTS AND PUBLICATIONS
1894	Resigned from U.S. Geological Survey; In retirement devoted his time largely to studies in ethnology and philosophy			
1895			Wrote 'Physiographic processes', first monograph of the National Geographic Society	
1896	With his wife, went to live in a cottage at Haven, Maine			
1898			*Truth and error*	
1900		Field study of the Carab and Arawak Indians; visits to Cuba and Jamaica	'The lessons of folklore'	
1902	Died 23 September at Haven, Maine			

Elisée Reclus
1830-1905

BÉATRICE GIBLIN

Élisée Reclus had his own significance among the geographers of his time, though not as a world explorer like his predecessors, nor as a renowned academic like Vidal de la Blache, for he was a teacher only in the last years of his life, and then outside France. He is more widely known as an anarchist than as a geographer. His writing abounds in vivid description as well as encyclopaedic knowledge, and his greatness lies in the clarity of his work.

1. EDUCATION, LIFE AND WORK
Born in 1830 at St Foy-la-Grande, Élisée was the son of a Calvinist pastor of marked personality, who was far more concerned with his religious mission than his family responsibilities to his fourteen children, eleven of whom survived to adulthood. Two of the father's qualities which were markedly apparent in Élisée and his elder brother Élie (born 1827), were determination always to follow the dictates of conscience, and constant kindness to others. From his mother, a teacher, Élisée inherited his lucid style of writing. In 1848 he went to the Theology Faculty of Montauban, but he and his brother were both expelled a year later for their lack of work and their republican sympathies. Later they went to Germany and in 1851 became students at Berlin University where their courses included one by Carl Ritter on earth description. To supplement their income they gave lessons in French, but they refused to accept one pupil whose father held non-republican views.

Their convictions remained so strong that when the coup d'état of Napoleon III was successful in 1853, the brothers left France for England, where they met other exiles including Louis Blanc and Pierre Leroux. Élisée then decided to go to the United States as the land offering asylum to all political refugees, and for three years he acted as tutor to the children of a planter in New Orleans. There he abandoned all religious belief, disgusted by clergy who were also rich merchants, in some cases owning plantations and slaves. From this time his liberal ideological views gained strength. He had little opportunity for travelling, except for some trips on the Mississippi and one excursion to Chicago, and was tired of a sedentary life, so he went to New Grenada (Columbia). He was always eager to see new lands, and in a letter to his mother said that he felt the need to go and see the Cordillera, of which he had dreamed in his childhood and from which he was now separated only by the Gulf of Mexico. In New Grenada he tried to establish a farming enterprise that he had had in mind for some years, but he lost all his money and in 1857, penniless and in poor health, he was obliged to return to France. He settled in Paris in the hope of living as a geographer or journalist, and took with him his many notes for a book describing the world. This was to be his first geographical work.

Gradually he moved towards an anarchist outlook, fortified by the Protestant tenets of individual freedom to accept or reject all dogma and of the importance of personal morality. He held the view that liberty was always antithetical to authority and that immorality came from the denial of liberty. He supported Bakunin in the controversy with Marxism of 1872. This

was concerned largely with the power of the state,
for the aim of a Marxist revolution was to seize
political power and then break the state's bourgeois
administrative qualities, but an anarchist revolution,
inevitably inherently repressive, must destroy the
state itself.   This divergence of view emerged at the
5th Congress of the International Workers Association
held at The Hague in September 1872.

As a geographer Reclus was mainly self-taught
from reading and travel, for he spent only one year
as a student of Ritter.   On his return from America
he sent several articles to the secretary of the Paris
Geographical Society, Malte-Brun, son of the Danish
geographer who had written a *Universal geography* of
considerable renown in its day.   Sponsored by Malte-
Brun, Reclus became a member of the Société de Géographie in 1858, and was delighted to have the use of
its fine library.   Meanwhile he worked for Hachette,
the publishers, on their *Guides Joanne*.   Until 1870 he
travelled in France and other parts of Europe and
published a number of articles in the *Revue Germanique*, the *Revue des Deux Mondes*, the *Annales des
Voyages*, the *Bulletin de la Société de Géographie*,
and *La terre*.   This was, in Reclus's view, his life's
work, on which he had spent ten years of research and
writing.   The excellence of the material and the
clarity of the writing ensured success for *La terre*,
and it was re-issued twelve times, giving Reclus a
firm place among geographical scholars.   Another
work *L'histoire d'un ruisseau*, was published by
Hetzel, who also issued the works of Jules Verne.
This was less successful with the general public than
La terre, but the Ministère de l'Instruction Publique
and the city of Paris bought large quantities so that
the prices could be kept down.   This was certainly
an honour for a future *communard*.

Fame saved Reclus from deportation to New Caledonia for his active participation in the Commune of
Paris of 1871.   French and foreign scholars combined
to petition de Thiers in these terms: 'We think that
the work of this man belongs not only to his own
country but to the whole world and that if he is
silenced or condemned to exist far from civilization,
France will injure herself and diminish her rightful
place in the world.'   He went into exile, mainly
in Switzerland, until 1889 when he returned to Paris.
In 1892 the Paris Geographical Society gave him their
gold medal for his *Nouvelle géographie universelle*,
and in the following year he was similarly honoured
by the Royal Geographical Society in London.   Despite
his fine reputation in France, he never taught there
and in 1892 controversy arose over his possible
appointment to the staff of the Université Libre de
Bruxelles.   With other professors Reclus founded the
Université Libre Nouvelle, dedicated to the principle
of complete liberty in all its teaching.   The Belgian
government refused to allow students of this new
university to present themselves as candidates for
degrees, so naturally there were few students.   The
work, however, went on and Reclus continued teaching
in Belgium until his death in 1905.

2. SCIENTIFIC IDEAS AND GEOGRAPHICAL THOUGHT

Reclus's aim was to make geographical knowledge available not only to specialists but to a far wider public.

His success lay in his general scientific and theoretical contribution.

a. The scientific contribution

Reclus's first work, *La terre*, was in three parts;
Continents, Oceans, Life.   The descriptive chapters
contain no new scientific material, but they made
existing knowledge available to a wide public.   His
evocative and expressive style, with the innumerable
illustrations, made it possible for vast numbers to
imagine the diversity of places and of their physical
setting.   Although his work could be regarded as
'popular', scholars made good use of it: for example,
the geographer and geologist Emmanuel de Margerie spoke
with respect in the *Revue de Fribourg*, 1896, of 'what
is owed to Élisée Reclus, particularly on the study of
the earth'.   But de Margerie's book, *Les formes de
terrain*, of 1888, contained far more material of a
geomorphological character than Reclus's work, and the
rapid development from a descriptive physical geography
into a more scientific geomorphology partly explains
the rapidity with which the influence of Reclus declined.
Nevertheless Reclus had given geography scientific status
to a wide circle of readers, for before his day many
geographical works had been itineraries, guidebooks or
gazetteers.   It was also fortunate that Élisée Reclus
combined the precision of Joanne with the poetic
character of Michelet, notably in the *Nouvelle géographie universelle*.

To write, alone, a universal geography in nine-
teen volumes is a remarkable achievement.   Reclus was
not content merely to use the material available in
other geographies, but wherever possible added his
own observations in the field.   He was able to speak and
read several languages.   His literary skill made the
reader conscious of such varied features as the depth
of the equatorial forest, the savage power of a torrent
through a wadi, the massive sculpture of Norwegian
fiords.   His landscapes are alive: this is well shown
in his description of Manhattan, with its contrasts between the small houses and the skyscrapers, his pioneer
descriptions of the daily commuting, of the immense
activity in the city by day and its stillness by night,
and of the urban morphology with commercial, industrial
and varied residential quarters.   In such work he
was in advance of his time, though similar thought
reappeared in the work of Blanchard later on.

Reclus's *Universal geography* includes some fine
analytical work.   There is, for example, a penetrating
study of the varied effects of the British adminstration
of India, carefully considered later by Albert Demangeon
when he wrote his book on the British Empire.   Reclus
was not content merely to describe but wished also to
explain, and naturally his own political outlook affected
his treatment of history, economics and political systems.
His *Universal geography* had to be written as a geography
and not as an expression of his anarchist views (indeed
this was explicitly stated in the contract with
Hachette), and on occasion its author was unable to
express his whole mind.   It also had to be written regularly, as it was initially published in weekly parts.
This was one reason why it was quickly forgotten and eventually regarded, unjustly, as 'mere description'.   More justly
it has been criticized for an abundance of description and
for its somewhat monotonous structure dealing inexorably
with relief, climate, water resources, population and

regional study.

In the third work of his trilogy, *L'homme et la terre*, a far more authentic version of the thought of Reclus appears. The tone of this work led to considerable difficulties with the publishers, but in the end the book appeared. It differs substantially from the two earlier parts of the trilogy, and in effect is a complement to the *Universal geography*. It opens with a study of population in various countries of the world given in an historical context, for Reclus believed that the past shed light on the present. To some extent it was a social geography and inevitably this raised contention. 'It involves study of the struggle between classes, of social stability, and of the supreme power of individual decision: these three challenging topics must underlie any "laws" discerned in social geography; and in turn these laws must contribute much to the knowledge and influence which underlies the communal management of society, in harmonious relationship with the environmental influences already closely studied and indeed known. In defining the environment (*milieu*) of social geography there is a spatial element, with its thousand varied external characteristics but also the time element, with its constant transformation of circumstances, which in themselves have endless repercussions' (vol 1, 116).

All this means that determinism is discarded; for identical physical characteristics do not result in identical economic and social responses. Although economic and social circumstances influence societies, it is economic and political systems that are responsible for the use of the environment. These were a source of devoted interest to Reclus, expressed in such studies as those of migration movements, economic and cultural relations, political revolutions, social transformations, even those of family life, land ownership and religious phenomena.

In the last volume of *L'homme et la terre* Reclus gives his own view of world society. He emphasizes the significance of the Industrial Revolution, especially in its spatial aspects, including the concentration of population in certain areas and the associated rural exodus, the emergence of a wage-earning proletatiat and of a new urban morphology, and also the effects of financial speculation. He also analyzes the imperialism of Britain and the U.S.A. with an assessment of the strategic implications: he notes, for example, the significance to the British of Gibraltar as the entrance to the Mediterranean, and of Egypt on the route to India. As the control of India by the British developed, so the centre of power was moved (though it was not until 1912, after his death, that Delhi became the capital).

Social geography as seen by Reclus has not received the attention in France that might have been expected. Many French geographers were more concerned with regional aspects, and particularly with the permanence of regional units, than with their modification with time. Study of the countryside was favoured rather than study of towns, where the changes due to the Industrial Revolution were more apparent and indeed challenging. The strength of Reclus's work as a geographer lies partly in its relationship to sociology, economics and anthropology and in his ability to recognize the complexity and dynamic qualities of human situations.

b. *Geographical ideas*

The geographer must be an observer of the natural world, as was Reclus himself. This natural world is the end product, at any one time, of an evolution which is continuous. In it man is a physical reality rather than a metaphysical being, and the natural order cannot be explained in metaphysical terms but only in scientific terms. Science inevitably seeks the truth by the study of nature, using the experience of earlier scholars and working on hypotheses which are capable of investigation. Geographers can only reach conclusions on a basis of observation and experience unimpeded by acquired misconceptions based on tradition. Observation must lead to a true explanation of facts. One area differs from another and resources are unequally distributed, though in the world as a whole there is enough for everyone. Reclus thought that the study of geography was a vindication of anarchism, for it proved that the anarchist ideal of 'bread for all' was not a Utopian dream but a possibility. Further, by geographical observation man could discover the 'natural' organization of space appropriate for an anarchist society formed of local communes adjusted to the local environmental resources, in which people co-operate with one another in concerns relative to the welfare of all.

Atheistic convictions meant that Reclus rejected all supernatural views of the world: in *L'homme et la terre*, (vol 6, p 389), he said that 'religion ascribes to God the revelation of all truth while science, having broken the bonds which tied Man to the unknown, seeks truth only in the observation of Nature.' Only science can reveal and explain the natural order, and geography is a science that studies nature objectively and guides the adaptation of society to the natural world. A deep knowledge of the reactions seen in nature can make possible, indeed guarantee, an equable utilization of resources and avoid supposed progress, which in time proved destructive of the earth. Reclus thought that there could be a *harmonie secrète* between the earth and its people, but this was to be discerned not by following determinism but by studying the interactions of the human and the physical. His work is in some aspects comparable with that of ecologists of the present time, but in his social geography he also showed with scientific argument that multitudes were oppressed by certain social systems and by political authority. These resulted in monopoly profits and in the acquisition of space by a few against the interests of the many. As he conceived social geography, it completely justified his anarchist outlook.

History and geography were linked in nineteenth-century France, with geography as the lesser relation. This was not the view of Reclus for, following Michelet, he held the opinion that if all history began as geography, geography becomes history in time through the continual interaction of individuals. He synthesized this idea on the title page of *L'homme et la terre*: 'La géographie n'est autre chose que l'histoire dans l'espace, de même que l'histoire est la géographie dans le temps.' It is not the prime objective of geography to provide maps of provinces or states for historians, nor to learn from them how historical events are influenced by the environment, but rather to present and explain the organization of the world. Vidal de la Blache is often regarded as the first geographer to integrate historical explanations with geographical reasoning, but clearly he had a precursor in Élisée

Reclus who, incidentally, was strongly aware of his own contemporary history.

Geography was 'global' in Reclus's work, concerned with the relation between human groups and the environment, and he did not favour its division into specialisms such as topography, climatology and demography. Geography must begin as a whole, including cosmography, natural history, history and topography, for the natural world with which it deals is an immense synthesis to be seen in its totality and not piecemeal. Reclus opposed Drapeyron's view that all geographical study, at whatever level, should begin with topography for, as he said in one of his letters, such a view was narrow: the greater purpose of the teacher was to show that everything was part of an infinitely greater whole within which endless variations might be sensed.

Reclus was always concerned about teaching geography effectively, for he wanted to make it a living subject for his pupils. He was deeply critical of many contemporary textbooks, as they presented a world in which science had done its work, with approved conclusions neatly paraphrased and presented, with a certainty characteristic of a religion that had in fact become superstition. He criticized the popular manuals of Levasseur as 'too dry', and thought that those of Cortambert were virtually devoid of human feeling. The earth seemed to be made of metal, the towns of papier-mâché and the men of cardboard.

Cartography remained an abiding interest, and he gave deep consideration to the maps in the *Universal geography*. He was fortunate to have the services of an old friend and comrade in the anarchist movement, Charles Perron, in the preparation of several thousand maps. He also conceived, in 1895, the idea of a vast globe, (2.4m in diameter) on the scale of 1:1,000,000 to demonstrate the relief of the world. Shown at the Exhibition in 1900, this was a revolving globe to be made available for students and to be housed in halls in London, New York and Paris. Patrick Geddes favoured the scheme, but it was estimated that each globe would cost £200,000 and no appropriate benefactor could be found.

3. INFLUENCE AND SPREAD OF IDEAS

As a member of the Paris Geographical Society Reclus was in touch with French geographers, explorers and geologists until 1870, after which he maintained a constant correspondence with them. Several of them gave him considerable help with his *Universal geography*, in which he made frequent mention of Duveyrier, the explorer, and also Franz Schrader, who was responsible for the cartography in several of the volumes and also supervised the writing of the chapters on the Pyrenees. Reclus's international fame enabled him to meet many geographers, historians and sociologists from other countries, some of whom gave him considerable help with maps and documentary sources. At the end of each volume Reclus lists the major authorities who had helped him, for example Glaziou and de Paranague of Brazil, Moreno of the Argentine, Vergara y Velasco of Colombia, Redway of U.S.A. Kuyper of the Netherlands, de Foucauld in Morocco, Malatesta in Italy, de Gerando in Hungary.

Kropotkin, the geographer and anarchist, was in close touch with Reclus. They worked together on the volumes of the *Universal geography* which dealt with northern Europe and Russia. Reclus favoured the ideas of Charles Darwin, and also the sociological principles of Auguste Comte who thought that sociology should seek to establish laws, particularly of an historical character. He also favoured the work of Herbert Spencer, the English sociologist. Though Reclus's work was in tune with that of advanced thinkers of the time, it does not appear to have been influential among sociologists or economists, many of whom found his views similar to those of the Marxists.

Nevertheless, the influence of Reclus is undeniable. The last volumes of the *Universal geography* were warmly welcomed, notably in the *Annales de géographie* of 1893 where the volumes on South America were hailed as among the best of the whole series, as well furnished with information and graced by elegant writing as any of the others. And in the *Revue de géographie* Drapeyron compared Reclus to Humboldt and Ritter. Another tribute to the *Universal geography* was its translation into English, Italian and Spanish and, for some of the volumes, into Russian. As noted earlier, *La terre* went through a dozen editions. The times were favourable. Geographical societies were becoming numerous; there was political, military and economic penetration of new territories; and public curiosity about new lands was immense.

In assessing Reclus's contribution to geography, some commentators were strongly aware of the problem of separating the anarchism from his work. An obituary of August 1905 in the *Revue de géographie* noted that he was obliged to live in exile for much of his life, mainly in Italy, Switzerland and Belgium. He was a man of generous nature with a strong faith in liberty and justice for all the world, devotedly serving those movements in which he believed. An appreciation in the *Temps* of 6 July 1905, the day after he died, spoke of his fine scholarship and of his unequivocal hope of establishing a Utopian society which led him to dangerous theories. It is unlikely that he did not link anarchism and geography in his mind; on the one hand, geography came from anarchism, for it revealed the mechanism of political and economic domination, and on the other hand, it seemed to Reclus that only a deep knowledge of natural laws and of the physical organization of space could make possible the formation of an anarchist society.

Without doubt the political aspects of Reclus's work have contributed to its current relative oblivion. But this is not the whole story. In the first place, Reclus was a teacher for only 11 years at the end of his life. He did not have the opportunity to make disciples, nor was he attached to any group of likeminded geographers. Secondly, he was not a theoretical geographer. Unlike Vidal de la Blache, he wrote no articles on the concepts, purpose, aims and methods of geography. The only indications of geographical theory appeared in his prefaces and his letters.

At the end of the nineteenth century scientific research in geography was developing mainly within the universities, and Vidal de la Blache became the founder of modern French geography. In human geography emphasis was placed on the characteristics of the physical environment, on the idea of relationship between its various features and on human groups as part of the general distribution of population. Social study was associated with ecological aspects and the *genres de vie* were considered not in sociological terms but in relation to the natural environment. The successors of Vidal de la Blache went further with these lines of research, especially in regional geography, and became 'essentially objective', but for that very reason limited in scope as they were so thoroughly concerned with space and distribution. The problems of life remained and were still, in the main, economic, social and political.

Even so, Reclus was not entirely forgotten. He is generally regarded as the last of the encyclopaedists, well versed in all aspects of geography. After his time a division between physical and human geographers developed. Reclus had no direct influence on school education, for he did not write textbooks, and his efforts to produce globes were less successful than Vidal de la Blache's efforts to produce wall maps, which became familiar to many generations of geography students even outside France. The modern growth of geography has been associated especially with the history of education, to which Vidal de la Blache (like many of his contemporaries in other countries) made a notable contribution. But this again is only part of the story, for without question Reclus remains, not only in his own time but even today, worthy of study as a geographer of scholarship and integrity.

The author has pleasure in acknowledging the help of Professor Gary S. Dunbar, of the University of California, Los Angeles, with the provision of bibliography and the checking of dates.

Bibliography and Sources

1. REFERENCES ON ÉLISÉE RECLUS

Regelsperger, S., 'Élisée Reclus', *Rev. Géogr.*, vol 55 (1905), 251

Schrader, F., 'Élisée Reclus', *La Géogr.*, vol 12 (1905), 816

Gallois, F., 'Élisée Reclus', *Ann. de Géogr.*, vol 14, 373-4

Kropotkin, P., 'Elisée Reclus', *Freedom*, vol 19, no 199 (1905), 6p.

Anon;, 'Elisée Reclus', *Geogr. J.*, vol 24 (1905), 337-43

Geddes, P., 'Élisée Reclus', *Scott. Geogr. Mag.*, vol 21 (1905), 548-55

Dunbar, G.S., 'Reclus, Élisée', in *Dict. Sci. Biog.*, vol 11, (1975), 337

Day Hem, 'Essai de bibliographie d'Élisée Reclus', *Pensée et Action*, (Paris, Brussels, 1956), 26 p.

Reclus, P., *Les frères Élie et Élisée Reclus ou du protestantisme à l'anarchisme*, Paris, 1964, 210p.

Giblin, B., *Élisée Reclus, pour une géographie*, thesis University of Paris, VIII (1971), 250p.

Maitron, J., 'Bibliographie des publications anarchistes d'Élisée Reclus', in *Le mouvement anarchiste en France de 1880 à nos jours*, Paris, 2 vols, (1975), vol II, 405-07

The letters of Reclus were published as *Correspondance*, 3 vols (1911-25), 352p., 519p., 339p.

2. SELECTIVE AND THEMATIC BIBLIOGRAPHY OF ÉLISÉE RECLUS

a. Physical geography

1859 'Étude sur les fleuves', *Bull. Soc. Géogr.*, sér 4, vol 18, 69-104

1859 'Préambule à 'De la Configuration des Continents' de Carl Ritter', *Rev. Germ.*, vol 8, 241-67

1862-4 'Le littoral de la France', *Rev. Deux Mondes*, 15 Dec. 1862, 901-36; 1 Aug. 1863, 673-702; 15 Nov. 1863, 460-91; 1 Sept. 1864, 191-217

1865 'Les oscillations du sol terrestre', *Rev. Deux Mondes*, 1 Jan., 57-84

1865 'Études sur les dunes', *Bull. Soc. Géogr.*, vol 9, 193-221

1867 'L'Océan, étude de physique maritime', *Rev. Deux Mondes*, 15 Aug., 963-93

1868 *La terre. Description des phénomènes de la vie du globe*, Paris, 2 vols, 823p., 781p.

1873 'Les pluies de la Suisse', *Bull. Soc. Géogr.*, sér VI, vol 5, 88-91

1873 'Note relative à l'histoire de la mer d'Aral', *Bull. Soc. Géogr.*, sér 6, vol 6, 113-18

1908 *Les volcans de la terre*, Société Belge d'Astronomie, 515p. (posthumous)

b. Regional geography

1859 'Quelques mots sur la Nouvelle-Grenade', *Bull. Soc. Géogr.*, sér 4, vol 17, 11-41

1861 *Voyages à la Sierra Nevada de Ste Marthe. Paysage de la vie tropicale*, Paris, 308p. 2nd ed 1881

1874 'Voyage aux régions minières de la Transylvanie occidentale', *Tour du Monde*, Paris, 704-06

1876-94 *Nouvelle géographie universelle. La terre et les hommes*, Paris, 19 vols

1902 *L'empire du milieu. Le climat, le sol, les races, la richesse de la Chine* (with Onésime Reclus), Paris, 667p.

1905 *Nouvelle introduction au dictionnaire géographique et administratif de la France*, 7 vols, (1890-1905).
The introduction is in a special volume, 163p.

c. Social and political geography

1862 'Le coton et la crise américaine', *Rev. Deux Mondes*, 1 Jan., 176-208

1863 'Les Noirs américains depuis la guerre civile', *Rev. Deux Mondes*, Mar., 315-40

1866 'Les républiques de l'Amérique du Sud, leurs guerres et leur projet de fédération', *Rev. Deux Mondes*, Oct., 953-80

1867 'Les Basques. Un peuple qui s'en va', *Rev.
Deux Mondes*, mar., 315-40
1868 'L'election présidentielle de la Plata et la
guerre du Paraguay', *Rev. Deux Mondes*, 15 Aug.,
891-910
1894 'Hégémonie de l'Europe', *La Soc. Nouv.*, 10 année,
vol 1, no 112, Apr., 433-43
1903 'Le panslavisme et l'unité russe', *Rev. Deux
Mondes*, 1 Nov., 273-84
1905-08 *L'homme et la terre*, Paris, 6 vols

d. Tourism
1860 *Guide du voyageur à Londres*, in *Collection Guide
Joanne*, Paris, 530p.
1862 *Itinéraire descriptif et historique de l'Allemagne*
(with G. Hickel) 2nd ed,
1862 *Allemagne de Nord*, in *Collection Guide Joanne*,
Paris, 717p.
1862 Introduction to *Itinéraire général de la France,
les Pyrénées et le résaux des chemins de fer du
Midi et des Pyrénées*, in *Collection Guide Joanne*,
Paris, pp. i-lxxii
1864 *Stations d'hiver de la Méditerranée: Nice, Hyères,
Cannes, Monaco, Menton, Bordighera, Sanremo* (with
P. Joanne), Paris, 383p.

e. Anarchism
1880 *Évolution et révolution, publication des Temps
Nouv.*, no 38, 25p.
1885 Préface to *Paroles d'un révolté*, Paris, pp.i-x
1890 *Richesse et misère, Publication de la Révolte*,
2nd ed, 72p.
1892 Prefaces to M. Bakunin, *Dieu et l'état*, Paris,
pp.i-ix, and to P. Kropotkin, *La Conquête du pain*,
Paris, pp. v-xv
1892 *L'évolution, la révolution et l'idéal anarchiques*,
Bibl, Sociol., Paris, no 19, 296p.
1895 *L'évolution légale et l'anarchie*, Brussels, *Bibl.
Temps Nouv.*, no 3, 17p.
1900 'Les colonies anarchistes', *Temps Nouv.*
7-13 July, 4p.

f. Other works
1864 'De l'action humaine sur la géographie physique.
L'homme et la nature', (review of *Man and nature*
by G.P. Marsh, *Rev. Deux Mondes*, 1 Dec. 762-71
1865 'L'histoire du peuple americain' (review of work
by A. Carlier), *Bull. Soc. Géogr.*, sér 5, vol 9,
143-64
1866 'Du sentiment de la nature dans les sociétés
modernes', *Rev. Deux Mondes*, May, 352-81
1869 *Histoire d'un ruisseau*, Paris, 320p.
1880 *Histoire d'un montagne*, Paris, 255p.
1894 'Leçon d'ouverture du cours de géographie comparée
dans l'espace et dans le temps', Brussels, 16p.
1895 *Projet de construction d'un globe terrestre à
l'échelle du cent millième*, *La Soc. Nouv.*, 16p.
1901 *L'enseignement de la géographie, globes, disques
globulaires et reliefs*, 2nd ed, 1902, Brussels, 12p

3. UNPUBLISHED SOURCES ON ÉLISÉE RECLUS
A bibliography of the works on geography and sociology
of Élisée Reclus, almost certainly by his sister Louise
Dumesnil, is in the Bibliothèque Nationale, Paris. Also
in the library's manuscript collection are manuscripts
and letters by Reclus.

*Mme Béatrice Giblin, Agrégée de l'Université, is
Assistante de géographie à l'Université de Paris VIII.
Translated by T.W. Freeman.*

DATES	LIFE AND CAREER	ACTIVITIES, TRAVEL, FIELDWORK	PUBLICATIONS	CONTEMPORARY EVENTS AND PUBLICATIONS
1830	Born 5 March at Ste Foy-la-Grande			
1836				Translation into French of part of Carl Ritter's *Erdkunde* as *Géographie générale comparée*
1848	Passed school-leaving examination			
1851	Became student of Carl Ritter	Refugee in England – also visited Ireland		
1853		Settled in New Orleans		Coup d'état of Napoleon III
1856		Visit to New Grenada (Columbia)		
1857	Returned to France	Began work for Hachette firm		
1858	Joined Paris Société de Géographie			Last volume of von Humboldt's *Kosmos* published
1862	Met Proudhon and Blanqui			
1868			*La terre, description des phénomènes de la vie du globe*	
1869			*Histoire d'un ruisseau*	
1871	Arrested and imprisoned for his participation in Commune activities			Paris Commune
1872	Exile in Switzerland			Levasseur, *Étude et enseignement de la géographie*
1873		Travelled in Austria		
1874				Charles Darwin's *Descent of Man* (2 vols 1871) published in a French translation
1876			*Nouvelle géographie universelle*, vol 1	
1877	Met P. Kropotkin			
1880			*Histoire d'un montagne*	
1882		Visited Hungary and Asia Minor		F. Ratzel, *Anthropologie*, vol 1
1884		Travelled in Egypt, Tunisia and Algeria		Translated version of E. Haeckel' *Natural history of the Creation* (1868 appeared as *Histoire de la création des êtres organisés*)
1885		Toured Spain and Portugal		

DATES	LIFE AND CAREER	ACTIVITIES, TRAVEL. FIELDWORK	PUBLICATIONS	CONTEMPORARY EVENTS AND PUBLICATIONS
1886		Visited Italy		
1889	Returned to Paris	Went to Canada and United States		Vidal de la Blache, *États et nations (autour de la France)*
1891		Second visit to Canada and United States		F. Ratzel, *Antropologie*, vol 2
1892	Gold medal, Geographical Society of Paris	Travelled in Spain and Portugal	*Évolution, révolution et idéal anarchique*	
1893	Gold medal, Royal Geographical Society, London	Toured Brazil, Uruguay, Argentina, Chile		
1894	Professor, Université libre de Bruxelles			Vidal de la Blache, *Atlas-Général Histoire et Géographie*
1895			*Évolution légale et l'anarchie*	Ratzel, *Politische Geographie*
1903				Vidal de la Blache, *Tableau de la géographie de la France* (as vol 1 of Levisse, E., *Histoire de France*)
1905	Died 4 July at Ixelles, Belgium		*L'homme et la terre,* 6 vols New introduction to *Dictionnaire géographique et administratif de la France,* Joanne, 7 vols	Vidal de la Blache, 'La conception actuelle de l'enseignement de la géographie', *Ann. Géogr.,* vol 14 (1905), 193-207

Nathaniel Southgate Shaler

1841–1906

WILLIAM A. KOELSCH

Nathaniel Southgate Shaler has been described
as 'a geologist by training and yet a geographer in
instinct'; during his lifetime he also made contributions to history, literature, social philosophy, and
popular science.   Primarily through his prolific
writing and popular lecturing on the relations of man
and the earth, he may be seen as a principal, if
neglected, link in the chain of American thought on
human geography connecting George Perkins Marsh with
the environmental consciousness of the late twentieth
century.

1. EDUCATION, LIFE AND WORK

Shaler was born into an upper-class slaveholding
Kentucky family in 1841.   He grew up with the normal
rural boy's interest in wild and domestic creatures,
and had developed a taste for geology and natural
history by the age of twelve.   In 1858 the seventeen-
year-old Shaler, convinced that he was 'more naturalist than humanist', entered Louis Agassiz's laboratory
in Harvard's Lawrence Scientific School, where he
worked for three years among a brilliant group of
naturalists.   His concentration on geology rather
than zoology made him unique among Agassiz's pupils.
Although 'the Master' set him to work on problems of
paleontology, his interests in the earth's surface
were nurtured through his scientific associations in
the Boston Society of Natural History, and through
fieldwork with such geological researchers as
Charles T. Jackson, Jules Marcou, and Henry and
William Barton Rogers.   In the summer of 1861 Shaler
went with two fellow students on a four month
expedition to the Gulf of St Lawrence in search of

Silurian fossils, a trip important in developing
a major scientific interest in the zone of contact
between land and sea.   In the field, he came to
a personal 'sense of the actuality of the earth
which,' he said, 'has served me all my days'.

Following his return to Cambridge after two
years of military and civil service in Kentucky
during the Civil War, Shaler picked up the basic
instruction in zoology and geology in the Scientific
School from Agassiz's failing hands.   At the age
of 28 he was appointed Professor of Paleontology
(later of Geology) in Harvard College.   Shaler
also served for seven years as Director of the
second Kentucky Geological Survey, created several
summer field schools of applied geology for his
students (including William Morris Davis), and made
a large number of geological and geographical
investigations along the Atlantic seaboard for the
U.S. Coast Survey and for the U.S. Geological Survey.
He also worked as a mining consultant and held a number
of appointments on Massachusetts state scientific
commissions.

Travel in search of health or intellectual
stimulus, which took him to England and to most of
Western and Mediterranean Europe, brought him into
contact with such leading scientific men as Charles
Darwin (whose theories he had adopted, contrary to
Agassiz's teachings), T.H. Huxley, Sir Francis
Galton and Léonce Élie de Beaumont.   Through travel
he increased his familiarity with and interest in
classic problems of structural geology, and also
intensified his sense of the aesthetics and significance of the designed or humanized landscape, a
theme on which he was to write and also to lecture

in the Harvard School of Architecture.   After 1888
his love of contact with and care for the face of
the earth found small-scale expression in 'Seven
Gates', a rural retreat on his beloved Martha's
Vineyard.   Plans for a productive retirement
there were shattered by an unexpected illness and
death during his planned last semester as Dean of
the Lawrence Scientific School.

2. SCIENTIFIC IDEAS AND GEOGRAPHICAL THOUGHT

Shaler's writings touch on at least three major
concerns of the human geographer, the character
and use of coast and inundated lands (especially along
the Atlantic seaboard of North America), the ethics
and aesthetics of resource management and conservation, and the relations of man and environment in
past times.   Many of his early geological studies
were concerned with features of the shoreline and
with certain offshore islands of New England.   These
and other studies of man's oceans and shores also
found expression in a popular book, *Sea and land*.
Beginning in the 1880s, Shaler engaged in extensive,
pioneering studies of the unutilized coastal and
fresh-water swamps and marshes of the Atlantic shore,
particularly concerning the possibility of their
reclamation for agricultural purposes.   His work
in this area is the eastern counterpart of J.W.
Powell's appraisals of the agricultural potential
of the arid lands west of the 100th meridian.

Shaler also made many contributions to the
nation's inventory of natural resources and to
the assessment of man's demands on them, beginning
with his direction of the comprehensive and diversi-
fied resources survey of his native Kentucky.   His
private consulting work, as well as his state and
federal assignments, led him to studies of the
alteration of the forests, of mineral depletion,
of unexploited energy sources, and of mineral
recovery possibilities.   In the late 1880s he
became interested in the processes of soil formation
and depletion, and his 1891 monograph on 'The origin
and nature of soils' has been called by a modern
scholar, Clarence Glacken, 'a landmark in the history
of soil concepts'.

Shaler was an early advocate of government
policies in both soil and forest conservation, and
wrote to raise the public's consciousness of these
resources as a trust for coming generations.   He
also advocated the protection of wild species and the
public reservation of areas of natural beauty in the
Eastern U.S., such as the Appalachian National Park.
The culmination of this line of research and writing
came in his philosophical and conservationist
synthesis, *Man and the earth*.   Here he stressed the
effect of economic and population development on the
energy and food resource base as well as the possibility of solving growth problems through human
intelligence and technology. But he also included
essays on 'The beauty of the earth' and 'The
attitude of man to the earth'.   Many of his specific
ideas, such as the potential of solar energy, were
ahead of their time.

Shaler's third main contribution to American
human geography is found scattered through his
writings on the relation of environment and man in
past times.   In a series of articles and books from
1880 to 1896 he wrote much on the mutual relations of
nature and man in America, among them one of his
best known (and most deterministic) volumes, *Nature
and man in America*.   In this and other studies
Shaler considered such culturogeographic problems as
the eastward range of the bison, the transfer of Old
World domesticated animals and plants and the impact
of New World additions, the concept of climatic
optimum, the cultural appraisal of prairies, the
'Great American desert' and other geographical myths,
cultural aversion to certain kinds of animal flesh,
folk tillage practices and the impact of agricultural
innovations, the use of anthropometric and actuarial
data to measure population quality, and the history
of aesthetic responses to landscape.   His treatment
of these and other problems was sporadic, and in
many cases has been superseded; his importance is
that he raised these issues before a wide audience
and, in a sense, set a long-hidden agenda for
twentieth-century human geography.

Like most American evolutionists of his generation, Shaler is not always consistent in his interpretation of the relation of man and nature ('consistency was no bugbear to his large mind', said Davis).
As an historical-cultural geographer he falls sometimes in the Marsh-Vidal-Bowman-Sauer tradition and
at others in a more determinist line of descent after
the manner of Guyot Buckle or Semple.   Something of
neo-Lamarckian in his views of race, he originally
entertained the notion that fixed racial characteristics had been given to each 'folk' by the environment of its original habitat, and modified only
slowly (especially through the range of environment-
ally permitted occupations) even after migrations.
Yet there was, for Shaler, a scale of cultural
advancement from 'savagery' to 'civilization', and
clearly on this scale the English-speaking peoples
had evolved to the point where they could and did
modify their environments substantially.   In his
studies of man's activity in specific areas, in his
post-1900 writings on social philosophy, and in his
evolving faith in the possibility of further modification of the landscape through the constructive
application of modern scientific knowledge, Shaler
could be both possibilist and prophet of a new
earthly creation.

Shaler's view of man's place in the natural
order was in many ways a latter-day extension of that
of Marsh, whose work he admired and on whose 'great
masterpiece', *Man and nature,* he patterned his *Man
and the earth*.   The Marshian mode is also apparent in
number of his other works, particularly the chapters
on forests and soils in *Aspects of nature*.   The
two men shared a concern with several major themes
of cultural geography: man's modification of vegetation by fire; the domestication of animals and
plants; the erosion of soils and the loss of what
Shaler called 'tillage values'; the aesthetic
appreciation of the humanized landscape; and most
clearly the central role of man ('this innovating
creature', in Shaler's words) as a shaper of the
landscape and as an appropriator of Nature's bounty.

Shaler lacked the element of romantic primitivism
characteristic of Marsh's generation and many others
since.   He welcomed the prospect of a thoroughly

domesticated landscape, 'an intensely humanized earth, so arranged as to afford a living to the largest possible number of men'. Man is not, in Shaler's view, a trapped animal; he is a thinking being, part of whose claim to greatness must be in his wise trusteeship of the earth he transforms by the mere act of living on it. In harmony with his close friend the philosopher William James, Shaler accepted the hypothesis of an open universe, and stressed man's creative intelligence operating through scientific awareness to solve the problems of existing together on this planet.

3. INFLUENCE AND SPREAD OF IDEAS

His fellow geologists elected Shaler President of the Geological Society of America and in 1903 ranked him eighth among living American geologists. The very diversity of his interests, however, as well as his institutional roles, prevented Shaler from producing major finished research contributions to human geography on the order of his geological studies, or from founding a 'school' of human and regional geography after the manner of his French contemporary Paul Vidal de la Blache. Shaler had a formidable reputation as an inspiring, if not especially rigorous, teacher of undergraduates, in whom he tried to encourage 'a broader and more humane outlook on the world'; for many years almost every Harvard College student enrolled in 'Geology 4'. His limited advanced teaching, however, was largely in paleontology and field geology and often included taking advanced geology students on governmental assignments. Much of the rest of his time and limited energies was absorbed by administration. Thus he had little opportunity, had he the inclination, to build up a corps of advanced students in human geography.

It was Davis and his enthusiastic disciples who set the paradigm for the next generation of American geographers, and Davisian science was not inclusive enough to permit the kind of human geography adumbrated by Marsh and Shaler to flourish at Harvard or in the profession. Of the many major figures of early 20th-century American geography who passed through Shaler's Harvard, only A.P. Brigham (whose *Geographic Influences in American History* incorporates several of Shaler's ideas), Ralph S. Tarr (of whose *Elementary Physical Geography* the same is true) and, less clearly, C.F. Marbut (whose interest in soil was awakened by Shaler, but whose major concepts were derived from Russian soil scientists), worked even in part in the Shaler tradition. It was left to J. Russell Smith, a Pennsylvania Ph.D., to hail Shaler as 'my unseen master' and term him, accurately, as 'a pioneer appreciator of the relationship between earth and man'.

Shaler's unique gifts as a writer and lecturer, plus the visibility of his Harvard post, made him the foremost American popularizer of late Victorian earth science. He was prolific (his bibliography runs to some twenty books and two hundred articles, papers and reports) and widely read in such leading journals of opinion as *The Atlantic Monthly*, or in the collections of these articles and popular versions of his geological survey reports published between

hard covers. His writing habits reflected both economic necessity and his commitment to bringing science closer to general American thought, but his intellectual style was basically that of the lecturer-essayist. He had matured in an age and setting which still regarded literary attainment as the highest earthly good, and wrote first of all 'to get his own bearings and to reveal to others something of the vastness spreading before him'. Much of his writing was in the form of speculative exploration, and in the end, his ideas on geographical themes were neither developed in the fashion required by the newer monographic modes of scientific expression nor institutionalized in places where his insights might have been more systematically explored and criticized, either by himself or others.

Davis's estimate of him as 'a man whom students flocked to hear because of the way he taught rather than what he taught' has stood as the definitive judgement on Shaler's life and work. With the end of his exciting lectures and of his reading public, Shaler's reputation declined and his geographical writings were ignored. In the mid-1950s, however, references to his *Nature and man in America*, *Man and the earth*, and his writings on soils began to appear in the literature of American geography, particularly through recognition of his and Marsh's ideas in an interdisciplinary symposium on 'Man's role in changing the face of the earth'. American intellectual historians also began to examine him as a commentator on race, a purveyor of science to the upper and middle classes, a pioneer conservationist, and a student of the frontier process; one student, Walter Berg, has called him 'the leading exponent of the geographical interpretation of American history' in the 1880s and 1890s.

With the more recently renewed interest among American geographers in the aesthetics of the humanized landscape, in the intellectual antecedents of modern geographical science, in the limits of growth, and in the ethic of 'reconciliation with the environment' which he advocated in *Man and the earth*, Nathaniel Southgate Shaler should once more find an audience. As one whose insights and suggestive hypotheses prefigured many of our current professional concerns and those of the educated public, we may expect that in the last quarter of the twentieth century he will come into his own as a trailblazer whose pathways are still worth climbing.

Bibliography and Sources

There are a number of contemporary death notices and
tributes by former students not listed here. Partial
bibliographies of Shaler's works are contained
in the first two titles listed.   See also the bibliographical references in Berg, below.

1. REFERENCES ON N.S. SHALER

Wolff, J.E., 'Memoir of Nathaniel Southgate Shaler',
  Bull. Geol. Soc. Am., vol 18, (1907), 592-609
Shaler, N.S. and S.P., *Autobiography of Nathaniel
  Southgate Shaler, with a supplementary memoir by
  his wife*, Boston, 1909
Davis, W.M. and Daly, R.A., 'Geography and Geology,
  1858-1928', in Morison, S.E. (ed), *The develop-
  ment of Harvard University, 1869-1929*, Harvard
  U. Press, Cambridge, 1930, 307-28
Malin, J.C., 'N.S. Shaler on the frontier concept and
  the grassland', in *Essays in historiography*,
  privately printed, Lawrence, Kans., 1953, 45-92
Berg, W.L., *Nathaniel S. Shaler: a critical study
  of an earth scientist*.   Unpublished Ph.D. Diss.,
  Univ. of Washington, Seattle, Bibliog., 1957
Haller, J.S., 'Nathaniel Southgate Shaler: a
  portrait of nineteenth-century academic
  thinking on race', *Essex Inst. Hist. Collections*,
  vol 107, no 2, April 1971, 173-93

2. PRINCIPAL WORKS ON N.S. SHALER

a. Man's coasts and inundated lands
1885 'Preliminary report on the seacoast swamps of the
  eastern United States', *U.S. Geol. Surv.*, *6th
  Ann. Rep.*, (1884-85), 353-98
1886 'The swamps of the United States', *Science*, 2nd
  ser, vol 7, (March 12 1886), 232-3
1890 'General account of the fresh-water morasses of
  the United States, with a description of the
  Dismal Swamp district of Virginia and North
  Carolina', *U.S. Geol. Surv.*, *10th Ann. Rep.*,
  (1888-89), 255-339
1893 'The geological history of harbors', *U.S. Geol.
  Surv.*, *13th Ann. Rep.*, (1891-92), part 2, 99-209
1894 *Sea and land: features of coasts and oceans with
  special reference to the life of man*, New York,
  xiii and 252p.
1895 'Evidence as to change of sea level', *Bull. Geol.
  Soc. Am.*, vol 6, 141-166
1895 'Beaches and tidal marshes of the Atlantic
  coast', *Natl. Geogr. Soc. Monogr.*, vol 1, no 5,
  American Book Co., New York, 137-68

b. Resource conservation and management
1875 'A state survey for Massachusetts', *Atl. Mon.*,
  vol 35, 357-63
1887 'Forests of North America', *Scribner's Mag.*,
  vol 1, 561-90
1888 'Animal Agency in soil working', *Pop. Sci. Mon.*,
  vol 32, 484-7
1889 *Aspects of the earth*, New York, xix and 344 p.
1891 'The origin and nature of soils', *U.S. Geol.
  Surv.*, *12th Ann. Rep.*, (1890-91), part 1,
  219-345
1895 'Preliminary report on the geology of the
  common roads of the United States', *U.S. Geol.
  Surv.*, *15th Ann. Rep.*, (1893-94), 259-306
1895 'Origin, distribution and commercial value of
  peat deposits', *U.S. Geol. Surv.*, *16th Ann. Rep.*
  (1894-95), 259-306
1896 'The economic aspects of soil erosion', *Natl.
  Geogr. Mag.*, vol 7, 328-38, 368-77
1896 *American Highways*, New York, xv and 293 p.
1901 'Proposed Appalachian Park', *N. Am.Rev.*, vol 173,
  724-81
1905 *Man and the earth*, New York, vi and 240p.

c. Man and environment
1877 'How to change the North American climate', *Atl.
  Mon.*, vol 40, 724-31
1885 *Kentucky: A pioneer commonwealth*, Boston, x and
  433 p.
1885 'Physiography of North America' and 'Effect of
  the physiography of North America on men of
  European origin', in J. Winson (ed), *Narrative
  and critical history of America*, 8 vols, Boston,
  vol 4, i-xxx
1890 'Nature and man in America', *Scribner's Mag.*,
  vol 8, 360-76, 473-84, 645-56
1890 'Science and the African problem', *Atl. Mon.*,
  vol 66, 36-45
1891 *Nature and man in America*, New York, xiv and
  240 p.
1892 *The story of our continent: a reader in the geography and geology of North America*, Boston,
  290 p.
1894 (editor and contributor), *The United States of
  America: a study of the American Commonwealth*,
  2 vols, (also issued in 3 vols), New York,
  1319 p.
1895 *Domesticated animals: their relation to man
  and to his advancement in civilization*, New York
  xi and 267 p.
1896 'Environment and man in New England', *N. Am.
  Rev.*, vol 162, 726-39
1898 'The landscape as a means of culture', *Atl. Mon.*
  vol 82, 777-85
1901 'American quality', *Int. Mon.*, 48-67

d. Miscellaneous writings
1889 'The geology of the island of Mount Desert,
  Maine', *U.S. Geol. Surv. 8th Ann. Rep.*, (1886-87
  987-1061
1893 *The interpretation of nature*, Boston, 305 p.
  (Winckley lectures)
1901 *The individual: a study of life and death*, New
  York, 351p.

1903 *Elizabeth of England: a dramatic romance*, 5 vols,
     Boston
1904 *The citizen: a study of the individual and
     government*, New York, 346 p.
1904 *The neighbor: the natural history of human
     contacts*, Boston 342 p.
1906 *From old fields: poems of the Civil War*, Boston,
     308 p.

e. Manuscripts
There is no major deposit of Shaler papers: some
materials are now known only from references in the
'Supplementary Memoir' (part of his autobiography
cited above).   The two largest remaining collections
are in the correspondence files of the U.S. Geological Survey in the National Archives, Washington,
D.C. and the latter repository also has a small
collection of materials concerning Shaler; there are
scattered letters and references in other Harvard
collections.

*William A. Koelsch is Professor of Geography at
Clark University, Worcester, Massachusetts, U.S.A.*

CHRONOLOGICAL TABLE: NATHANIEL SOUTHGATE SHALER

DATES	LIFE AND CAREER	ACTIVITIES, TRAVEL, FIELDWORK	PUBLICATIONS	CONTEMPORARY EVENTS AND PUBLICATIONS
1841	Born 20 February, Newport, Ky			
1859	Matriculated, Lawrence Scientific School, Harvard	Worked with Louis Agassiz, Museum of Comparative Zoology, Harvard (to 1862)		
1861		Anticosti expedition	*Lateral symmetry in brachiopoda* (thesis)	American Civil War (to 1865)
1862	S.B., *summa cum laude*. Married Sophia Penn Page. Commander, Fifth Kentucky Battery			G.P. Marsh, *Man and nature*
1864	Assistant in Paleontology and University Lecturer, Harvard			
1866		Field studies in Switzerland; travel and study in Italy, France, Germany (to 1868)		
1869	Professor of Paleontogy	First U.S. Coast Survey Appointment		C.W. Eliot inaugurated as President of Harvard
1873		Director, second geological survey of Kentucky (to 1880) Travel in England		Death of Agassiz
1875	Sc.D. in Natural History, Harvard	First summer field course in geology, Camp Harvard, Kentucky		
1881		Travel to England, France, Italy (to 1882)		
1884		Geologist-in-charge, Atlantic Division, U.S. Geol. Surv. (to 1900)	'Physiography of North America' in *Narrative and critical history*	
1885			*Kentucky*; 'Preliminary report on sea-coast swamps'	
1888	Professor of Geology	Lowell Lectures, 'Geographical conditions and life'		Geological Society of America founded
1891	Dean, Lawrence Scientific School (to 1906)	Winckley lectures, Andover Theological Seminary	*Nature and man in America*; 'The origin and nature of soils'	
1894			*The United States of America*; *Sea and land*	
1895		President, Geological Society of America	*Domesticated animals*	
1903	Ll.D., Harvard, 'Naturalist & humanist'		*Elizabeth of England* (drama)	
1904		Travel in Egypt, Greece, Italy	*The neighbour; the citizen*	Association of American Geographers inaugurated

DATES	LIFE AND CAREER	ACTIVITIES, TRAVEL, FIELDWORK	PUBLICATIONS	CONTEMPORARY EVENTS AND PUBLICATIONS
1905			*Man and the Earth*	
1906	Died 10 April in Cambridge, Mass.		*From old fields* (poetry)	Lawrence Scientific School discontinued
1909			*Autobiography of Nathaniel Southgate Shaler* published posthumously	

Thomas Griffith Taylor
1880–1963

J. M. POWELL

In the Moffett-Russell Collection in the National Library of Australia

Griffith Taylor was an Antarctic explorer, a senior physiographer and meteorologist with the Australian Federal Government, a distinguished university teacher and an indefatigable fieldworker. He studied geology in Sydney under Sir Edgeworth David and moved gradually towards geography before and during the First World War, principally through his association with David, with Scott in the Antarctic, and with W.M. Davis, A.J. Herbertson and Ellsworth Huntington. During the inter-war period, Taylor moved more boldly into the highly controversial application of geographical approaches to the study of settlement expansion. He founded Australia's first school of geography at Sydney University and after a brief but productive interval at Chicago founded Canada's first school of geography at Toronto. Although he is often remembered principally as one of the last modern exponents of 'environmental determinism', Taylor's prolific individualistic contributions towards physical, cultural, urban and political geography, together with his successful regional texts on Australia and Canada, reflect much more broadly the general scope and spirit of the great formative periods of university geography in the English-speaking world.

1. EDUCATION, LIFE AND WORK

Thomas Griffith Taylor, born 1 December 1880, at Walthamstow, Essex, was the first child of James Taylor, a highly qualified and well-travelled chemist and metallurgist, and Lily Agnes Griffiths. His parents had been born and raised in industrial Lan-

cashire, and their staunch non-conformism and teetotalism was inherited enthusiastically by their son. So too was James Taylor's career mobility and penchant for travel: he had been a research chemist on the Chilean nitrate fields before his marriage; in 1881 he became manager of a copper mine in Serbia; three years later he returned to Britain to take charge of the department of analytical chemistry at a major Sheffied steelworks, and in 1893 he moved to Australia as official Government Metallurgist in New South Wales. Dorothy, Griffith Taylor's sister and his faithful assistant throughout the early stages of his teaching career, was born in Serbia.

Taylor took his early schooling at a staid and unpretentious dame school, yet managed some early demonstrations of his independent streak by producing a few notes for the local newspaper before he was eight years old and a prize geographical essay which was published in the same newspaper before his eleventh birthday. In Sydney he was placed in élitist private schools, which he later declared responsible for his life-long dislike of sports and classics. At King's School Taylor established a close association with Frank Debenham, who would later join him in the Antarctic and subsequently become head of the geography department at Cambridge.

After a year as a clerk in the New South Wales Public Service, a choice of occupation which may have been conditioned by his father's prominent role or influence, and probably provided useful contacts and indicators of employment opportunities in later years, Taylor entered Sydney University in 1899.

Under the guidance of Professor (later Sir) Edgeworth David, he eventually decided to read science and mining engineering.   The emphasis in these courses on careful fieldwork and proficiency in mechanical drawing suited Taylor's own interests and abilities, and they were thereafter characteristic features in his idiosyncratic approach to his own teaching and writing. Edgeworth David was then engaged in pioneering work in Australian physiography and Taylor frequently accompanied him as an attentive and dedicated field assistant.   Douglas Mawson, another gifted student of David's, collaborated in some of this work; both men were deeply influenced by David's breadth of vision and particularly by his enthusiasm for Antarctic exploration, and Mawson chose to accompany David on Shackleton's expedition of 1907-09.   David persuaded Taylor to widen his interests by co-operating with Jose and Woolnough in producing an elementary geographical text on New South Wales for a series edited by the geologist J.W. Gregory, and he also encouraged his pupil's fascination for plaeontology, specifically for the Archaeo-cyanthinae, or limestone coral fossils, of the Flinders Ranges of South Australia.   The latter research provided the main task of Taylor's two years at Cambridge, the result of his graduation with distinction from Sydney and his receipt of a much-prized travelling scholarship.   Before leaving for Britain, Taylor was a demonstrator in geology in David's department and offered a short course in commercial geography, reputedly the first formal geography lectures in Australia.   By that time he had also exchanged correspondence with W.M. Davis and A.J. Herbertson.

While based at Cambridge between 1907 and 1910 Taylor enjoyed some association with the eminent geologists J.A. Marr and Sir Archibald Geikie. He also formed several strong friendships, notably with Raymond Priestley and Charles Wright (both of whom were later knighted), was elected a Fellow of the Geological Society, and accompanied Davis during his field investigation of glaciated topography in the Alps.   *Australia in its physiographic and economic aspects* (1911) was also completed at this time; it drew heavily on his Sydney lectures and owed a good deal to Herbertson's editorial amendments. Together with Debenham, Priestley and Wright, Taylor was invited to accompany Scott to the Antarctic.   Edgeworth David had already arranged for Taylor to join the new Commonwealth Weather Service on his return from Cambridge, and it was finally agreed that he should be permitted to act as the Weather Service's representative in Antarctica. Before joining Scott on the *Terra Nova* in November 1910, Taylor spent a very active period collating the meteorological data obtained by David and other members of Shackleton's party; he also assisted David in the preliminary mapping of the site of Australia's federal capital at Canberra and, with Debenham, Wright and Dorothy Taylor, studied glaciation in New Zealand's Mt Cook region.

Public interest in Scott's ill-fated expedition proved virtually insatiable and each of the survivors was in costant demand for many years; As senior geologist and leader of the western parties, Taylor was accepted as a major authority.   Specifically, he received the D.Sc. from Sydney University in 1916 for his Antarctic research; in addition, however, the mystique of the *Terra Nova* expedition undoubtedly boosted his own international career and, quite as importantly, the arduous field experiences made a deep and lasting impression on his youthful character.

Between 1912 and 1920 the confident and ambitious Taylor rapidly consolidated and extended his reputation in physical geography and related fields.   In his senior research position with the Weather Service he busily produced some of his most famous work on Australian meteorological conditions and 'climatic controls' in agriculture and settlement expansion. During the same period he received the Polar Medal, married Priestley's sister Doris in 1914, lectured brilliantly before the Royal Geographical Society and saw the publication of his popular account of Scott's expedition (1916).   The war years brought new demands and opportunities which greatly extended his narrow research interests into the area of development policy.   It is seldom realized, for example, that the staff of the Weather Service was then linked with the Intelligence Branch under the control of the Department of Home Affairs, and that the character of much of the work undertaken was dictated by considerations of national strategic concern, above all those of defence.   Taylor's early pronouncements on resource limitations and the inadvisability of tropical settlement have to be seen in this context; it is greatly to his credit that he established a firm independent viewpoint on these matters and refused to give way to the political and military arguments.   During this same period and for similar reasons he proposed the compilation of a national resources atlas -- the idea was taken up after the Second World War -- and served as a founder member of a national scientific advisory council, the forerunner of the present giant research body, the Commonwealth Scientific and Industrial Research Organization.   But his 'pessimistic' and 'unpatriotic' forecasts for Australia's development infuriated the settlement boosters, notably in Western Australia and Queensland, and Taylor found himself embroiled in bitter controversy at home and abroad.   Professor J.W. Gregory and other scholars of imperialist views publicly challenged his efforts to dilute the appeal of Australia as the major field for white immigration, and seized upon the 'environmentalist' theme to articulate their anger and distress.   The *Geographical Review*, through Isaiah Bowman, gladly published the products of Taylor's speculative research on settlement potentials and on conceptually related issues involving the influence of climate on racial origins after the *Geographical Journal* had brusquely rejected them.   Taylor soon allowed his membership of the Royal Geographical Society to lapse.

He usually enjoyed this increasing notoriety, but felt restricted by his official status. Assisted by David, he was a major contender for several academic positions, including the chairs of meteorology (at Sydney) and geology (at the University of Western Australia and at Otago University, in Dunedin, New Zealand); he also applied for the Readership at Oxford, vacant after Herbertson's death in 1915. Bowman, as Director of the American Geographical Society, invited Taylor to join the Society's research programme in Latin America, begun in 1920.

Professor David was then energetically pressing Taylor's case before the Sydney University authorities, who were contemplating the foundation of a geography department, and Taylor succeeded in using the American offer to further his claims. Characteristically, he fitted in an extensive tour of the East Indies before taking up his appointment as Associate Professor and foundation head of department in March 1921, with Dorothy Taylor as his assistant.

He was highly successful in building up the Sydney department and in keeping geography in the public eye. He never gave quarter in his frequent battles with the optimists in the daily press: indeed, his serious technical and academic approach was sometimes carelessly laced with seemingly obscure literary references and pedagogical rebukes, which his opponents considered arrogant and offensive. A few leading scientists and public figures declared their support for his lonely stand and his overseas contacts continued to encourage him. In 1923 he was awarded the Livingstone Centenary Medal of the American Geographical Society at the Pan-Pacific Science Congress in Sydney, but such honours did little to improve his position at home. The visit of the celebrated Arctic explorer Vilhjalmur Stefansson in the following year spurred on the boosters, for Stefansson enthusiastically proclaimed the 'myth' of hopeless aridity in a series of commissioned newspaper articles.

Taylor was then attempting to expand his Australian interests into broad global generalizations and in 1927 his *Environment and race* clearly established his name on the international scene. Advised by Huntington and Bowman, he accepted an invitation from Harlan Barrows to join the Chicago department, resigning from his Sydney post in 1928. Even then, however, he could not resist indulging in a few parting shots -- one of his favourite hobby-horses was the absurdity of the high status traditionally accorded to the study of classical languages, and his derisory remarks on this theme were widely covered in the Australian press, severely damaging his standing on campus and undoubtedly injuring the prospects of his successor. Yet geography in Australia had progressed considerably in his eight years of leadership at Sydney; he had become the founding President of the Geographical Society of New South Wales and the joint editor of its new journal, the *Australian Geographer*, which was first published in 1928.

Repeatedly denied the rank of full professor at Sydney, Taylor achieved that status on his arrival in Chicago in 1929, at the age of 48. Until 1935, when he moved to Toronto, he quickly consolidated his reputation as the leading authority on Australia while at the same time continuing to draw on his Antarctic experience and moving closer to the mainstream of the discipline. He made the acquaintance of Wallace Atwood, Arthur Compton, Mark Jefferson, Margaret Mead, Roderick Peattie, Robert Redfield, Ellen Churchill Semple, Paul Tillich and other major workers; and conducted field research in Europe and Latin America. In 1931 he was proposed by H.J. Mackinder as Vice-President of Session E of the British Association for the Advancement of Science. He enjoyed his years at Chicago professionally and

socially, but was far too 'British' in his orientation to consider it as a permanent base and never attempted to stake a personal claim to geographical analyses of the United States. In 1932, supported by F. Debenham, H.J. Mackinder, C.B. Fawcett and Sir Arthur Keith, he applied unsuccessfully for the Chair of Geography at Oxford; the attempt was not assisted by his continuing poor relationship with the Royal Geographical Society, which still had an official voice in the Oxford election. In 1935, shortly before his 55th birthday, he took up an appointment at Toronto University as Canada's first full professor of geography.

Geography was then fairly well established in Canadian schools and applied geographical techniques were employed in government circles. With the assistance of Donald Putnam and George Tatham, and the continued support of the economic historian Harold A. Innis, who had been largely responsible for his appointment, Taylor built a highly respected department with a strong research base. Once more, as in Australia, his creative personal style delivered invaluable pioneering statements, and again he vigorously declared that climate was the controlling factor in development. But his generalizations for Canada were, in fact, moderately optimistic and *Canada* (1947), a very marketable text, boldly summarized his challenge to the prevailing mood of pessimism. On the broader front, he continued the speculation on environmental influences in cultural and urban geography which he had initiated at Chicago. In *Environment and nation* (1936) he attempted a wild foray into the full sweep of European history and the more controlled *Urban geography* and *Geography in the twentieth century* appeared in the early post-war years. In 1938, sponsored by Debenham, he became President of Section E at the Cambridge meeting of the British Association for the Advancement of Science and was elected Vice-President of the Association of American Geographers at their Harvard meeting. In December 1940 he was unanimously elected the first non-American President of the Association of American Geographers. Immensely pleased with these high honours, he was nevertheless bitterly disappointed at his failure to achieve official recognition by the Royal Society in the late thirties and during the war years; there is some suggestion that this was due again, in part, to his rather aggressive posture in presenting his own claims as a potential F.R.S. In 1942 he was elected F.R.S.Can. but never fully participated in its affairs, allowing his subscription to lapse in 1954.

Taylor took up an invitation in 1948 from the Interim Council of the Australian National University to join a team of consultants advising on the establishment of research schools. He spent three months touring Australia and delighted in his unusually warm reception. Sponsored by the British Council, he also made a lecture tour of British universities in the company of his brother-in-law Sir Raymond Priestley, then Vice-Chancellor of Birmingham University, but the earlier tour of Australia was particularly heartening. He retired from Toronto University in 1951 with the title of Professor Emeritus, and before his departure he was elected Honorary President of the newly founded Canadian Association of Geographers, the professional organization which he had been promoting for several years.   He and his wife retired to a Sydney suburb and continued their active lives.   In 1954 he was made a Fellow of the Australian Academy of Science, the sole geographer in that fraternity, and in 1959 he became the first President of the Institute of Australian Geographers and was awarded an Honorary Doctorate of Letters by Sydney University.   He was still publishing, on Antarctica and on the contribution of geographical studies to world peace, in 1963, the year of his death.

2. *SCIENTIFIC IDEAS AND GEOGRAPHICAL THOUGHT*

a. *Scientific contribution*

In the days of his youth and early manhood, Griffith Taylor was constantly exposed to the influence of the kind of Social Darwinism ascribed to Herbert Spencer and his followers.   His forceful personality and early social and educational background were dominantly utilitarian and he was given a firm grounding in the physical sciences by David, quickly followed by close personal experience of the then dominant paradigm in physical geography, Davis's theory of stages in landscape evolution.   If it is admitted, in addition, that much of his most successful pioneering research was focused on continents brimming with environmental difficulties, his repeated efforts to seek 'Nature's Plan' and to establish 'underlying principles' and 'sequences' seem almost inevitable ingredients of his personal and professional credo. His early career as a scientist in the public service set him tasks which were incapable of solution from any narrow specialist viewpoint, while at the same time underlining the urgent need for scientific involvement in public affairs.   Ultimately he commanded the respect of scientists in general, not only in Australia, for promoting this viewpoint. When he migrated into academic geography, he made it his special task to identify and pursue a strong personal contribution towards the growth of the discipline, particularly by establishing and communicating its scientific basis.   'Environmentalism' was a useful organizing concept in the subject's formative years, and Taylor contributed handsomely in this respect.   His geological work had been very thorough and competent, but was essentially descriptive; apart from the fact that it helped to fill important information gaps, especially for Antarctica, its main value was in the practical training it afforded its author.   His adventurous forays in human geography were vastly more significant, despite their mixed reception.

   In his own self-image he was the enterprising frontiersman, pushing out one ragged and remote outer boundary after another, leaving the mundane infilling to more timid types.   Each speculative thrust served to confirm him in his conception of the role, yet other scholars diagnosed his condition as one of chronic pioneering, redolent with threats for himself and those who sought his company.   He was, above all, an individualist, making use of other workers' ideas when it suited him.   His early excursions into cultural geography, culminating in *Environment and race*

(1927), emerged from his own Australian experience, but were clearly heavily influenced in scope and approach by Huntington's deterministic ideas and by the palaeontologist/zoogeographer W.D. Matthew's *Climate and evolution* (1915), yet neither of these had attracted the favourable attention of leading anthropologists.   The formerly simplistic biological analogies had already been rejected or greatly modified by the social anthropologists and Taylor's enthusiastic use of hair texture and the cephalic index as a measure of racial groupings had been widely discredited by physical anthropologists Social Darwinism was also passé in biology and sociology, and scholars in each of these areas were intrigued at its apparent popularity in geography. Taylor's 'Zones and strata' theory (1936) was based on his crude classification, and never survived into the modern era.   Yet his theories on racial origins, distributions and migrations attracted considerable attention outside geography: *Environment and race* provided anthropology students with useful elementary geographical information and demonstrated the importance of mapping; it was widely read, sold well and was actually translated into Japanese and Chinese.   Neither his vague reply to the Germans' *Geopolitik* in his notions of 'Geopacifics', nor his efforts in urban geography, were deemed worthy of much academic attention within or beyond geography.

b. *Ideas held on geography and other sciences*

One of the primary attractions of Taylor's early geographical writings may have been that they demonstrated the essential indivisibility of the human and physical branches of the discipline while at the same time providing a 'deterministic' theme for human geography which suggested some potential for a badly needed scientific basis.   This is not to say that he was conscious of such a contribution, at least not in the beginning; he was certainly not philosophically inclined and cheerfully accepted the basic othodoxies of his day.   His uncritical acceptance of the Spencerian view was matched by the equally simplistic welcome he gave to Herbertson's notions on regionalization: he found these convenient, yet showed no essential grasp of Herbertson's insistence on man as an essential factor in the moulding of every region. During the 1920s, Barrows and Carl Sauer developed approaches for human geography which eschewed easy generalizations, emphasizing the mutual interdependence of man and environment, the need for detailed regional treatments and for close relationships with the social sciences.   In Europe, the French possibilists also rejected the environmentalist theme and stressed instead the scope of action available to man. Over the succeeding decades Taylor joined in the debate over determinism and possibilism, sometimes with annoying flippancy, without fully appreciating its philosophical heritage.

   Taylor's Presidential Address to the Australasian Association for the Advancement of Science (1923) provided his first major effort to define his personal view of geography, and his subsequent pronouncements were really only modifications of this. He emphasized the importance of preliminary and supportive training in the basic sciences, field and laboratory work, ethnology and the application of geographical approaches to national development. Geography was the study of the environment 'in its relation to human values', and together with biology and history it was an essential requirement in modern liberal education. The 'realm of geography' incorporated a variety of skills derived in increasing levels of importance from astronomy, geology, meteorology, biology, ethnology, economics and history in related sytematic approaches and in the investigation of the world's natural regions. His association with Barrows and acquaintance with the special 'liaison' role developed for geography by Chicago's innovative university administrators, assisted Taylor towards an appreciation of the complexities involved in the debate over the subject's role and status, and his inaugural address at Toronto in 1935, subsequently entitled 'Geography the correlative science', laid far greater stress on a grounding in related social and cultural disciplines than he had admitted in his Australian paper. His *Environment and nation* (1936) displayed his modified views very well, but its cavalier sorties into history and sociology were well received only by such fellow generalists as Toynbee, Huntington and Wells; Richard Hartshorne and other geography spokesmen sternly rebuked him. In his *Australia* (1940) he introduced the attractive idea of 'Stop-and-go determinism' to describe his own position -- man was the traffic controller altering the rate, not the direction, of progress. *Geography in the twentieth century* (1950), though a primary landmark in the history of the discipline, showed no new development in Taylor's own scrambled philosophy; the book's main value lies in the scholarly contributions he managed to attract as editor. In 1961 he chose to warn his colleagues once more about the dangers of underestimating the environmental controls -- 'man, to be successful, must study most carefully the environmental factors before embarking on the sole development which is likely to repay labours involved'. The integrating theme of 'environmental management' for geographical teaching and research in the 1970s is precisely the kind of development he wished to emphasize.

c. The world view

Taylor's remote colonial origins and his early preoccupation with mining geology and engineering seemed to give him sufficient excuse to parade his much-exaggerated philistinism towards art and music, but he travelled far more extensively and with far more purpose than most of his academic contemporaries, including those who entered geography from relatively more cultured backgrounds. One result of this was his strong stand against ignorant racism at whatever level he encountered it -- with the exception, in the early years, of some surprisingly naïve ideas about the American negro.

The purists inevitably condemned him for publishing too much too quickly. But Taylor firmly believed in the teaching-research-publication continuum; he was also sincerely concerned to function as an educator in the widest sense, and so he often aimed at producing books from his research whenever feasible, though he seldom completely disregarded the purification rituals of the established journals. His consuming search for broad generalizations continued to be well suited to this wider task of communication, although his professional colleagues were increasingly inclined to regard it as a very dated practice. As his contribution to the *New York Herald Tribune* Forum (1943) and his Messenger Lectures at Cornell University (1944) amply demonstrated, he responded as often as he could to invitations to make his ideas better known, and was always prepared to state his views simply and with conviction. In these respects he was well ahead of his contemporaries.

Towards the end of his life he attempted to convey his own sense of internationalism to his fellow geographers by calling for their urgent promotion of what he termed 'geopacifics', studies leading to global peace and understanding. This favoured theme entailed a frank and proper admission of the role of value judgements in the social sciences. Geographers were urged to identify desirable futures for society and to concentrate their efforts on directing their students to achieve those futures. In the same spirit, he made a well-publicized appeal to the Australian government to surrender its huge Antarctic Territory to the control of the United Nations.

3. INFLUENCE AND SPREAD OF IDEAS

Taylor was never so brazenly or completely deterministic as some authors have claimed. At first he derived some amusement from the title, but he moved rapidly to safeguard his considerable ego as the debate intensified, declaring that possibilism might be preferable in small area studies and especially in the world's better-favoured regions. Superficially, the whole issue may now seem an astonishing waste of time, but it had its place in tightening the emerging focus on epistemology and modes of explanation in geography. Though he maintained his view of himself as the *enfant terrible* until the end, there was commonly an eccentrically anachronistic tone to his work, not only in the later years. Somewhat unkindly, it might be said that the deterministic stance mainly provided a convenient model for his broad attacks on themes which interested him; similarly, his indulgence in controversial writing may have provided material rather than direction for the determinism-possibilism controversy.

It was Taylor's capacity for hard work and his maverick personality as much as any real intellectual achievements which made such a huge impact on the growth of geography during his lifetime. He was frequently as stimulating and provocative in his teaching as he was in his prodigious literary output, always seeking to communicate an independent interpretation, and highly inventive in his choice of illustrative material. His 'hythergraph', or 'climograph', while not entirely original, proved the most attractive and long-lived of his idiosyncratic illustrations, and his predilection for coining new terms left a distinctive legacy(*cf.* Stamp, L.D., *Glossary of geographical terms*, 1961). His insistence on fieldwork and laboratory techniques

helped for a time to reinforce the traditional
environmental bias in Canadian school geography,
and ensured that Australia and Canada followed
British rather than American approaches towards the
subject.   And for both Australia and Canada he bequeathed several highly original regional essays
which typified his explorer spirit, together with
two major regional texts, still widely recommended,
and fundamental introductory work in Australian meteorology used by Koppen in his famous *Handbuch* (1930).
Beyond these substantive regional texts and his early
scientific works dealing with Antarctica and
Australia, all of which gained full recognition for
their pioneering quality and basic competence,
Taylor's efforts to synthesize and theorize on the
grand scale evoked a mixed and largely temporary response in academic circles.   Within geography they
caused some confusion and embarrassment, yet in retro-
spect it may be admitted that they did much to encourage
others to examine a vastly extended range of content
and techniques.   His comparative insensitivity towards the complex epistemological arguments and
associated reorientations emerging after the Second
World War was true to character, and he was scarcely
in touch with geography's new preoccupations with
theory and methodology during the fifties and early
sixties.   But it is widely agreed that the general
nature and sheer quantity of his work over a very
lengthy period were more significant than the actual
calibre of his scholarship, and that the unique qualities of the man himself must stand out in very bold
profile in any lasting interpretation of the geographical profession over the past fifty years.

   Taylor believed that geographers and other
scientists had a duty to be professionally concerned
with the great issues of their day.   He sought out
those issues which were amenable to geographical analysis and did his best to clarify them for the educated
public as a whole.   In Australia during the twenties,
his presentation of his views on racial origins and
dispersals and on tropical development, were taken to
imply either depopulation or support for Asiatic peasant
settlement in Queensland and the Northern Territory,
and the powerful supporters of the White Australia
Policy howled him down.   His remarkably accurate
prediction of a population of about 19 millions by
the end of the century was treated with derision at a
time when the popular estimate ranged between 100
and 500 millions.   His role of honoured official
adviser during his brief return in 1948, was not
simply another instance of the renowned proclivity
of the Australians to reserve their recognition for
home-grown talent until the *imprimatur* of 'overseas'
success was displayed.   Taylor's pronouncements were
influential and very well-remembered in the political,
administrative and academic spheres (*cf.* Douglas,
1977), and in 1948 his rational approach obviously
appealed to those entrusted with the design of
regional and national development in the era of
reconstruction.   The present involvement of geographers in public policy in Australia, Canada and
elsewhere is the confirmation and extension of a
central philosophy which Griffith Taylor precociously
defined and proselytized.

Bibliography and Sources

1. REFERENCES ON T. GRIFFITH TAYLOR

Taylor, T. Griffith, *Journeyman Taylor. The education
    of a scientist* (abridged and edited by Alasdair
    A. MacGregor), London, (1958), 352 p.
Hartshorne, Richard, *Perspective on the nature of
    geography*, Chicago, (1959), 201 p.
Putnam, Donald F., 'Griffith Taylor, 1880-1963',
    Can. Geogr., vol 17, (1963), 197-200
Andrews, John, 'Griffith Taylor, 1880-1963', *Austr.
    Geogr. Stud.*, vol 2, (1964), 1-9
Aurousseau, Marcel, 'Obituary: Griffith Taylor,
    1880-1963', *Austr. Geogr.*, vol 9, (1964), 131-3
Browne, William R., 'Thomas Griffith Taylor', *Yearb.
    of Austr. Acad. Sci.*, Canberra, (1964), 41-53
Crone, G.R., 'Obituary: Thomas Griffith Taylor',
    Geogr. J., vol 130, (1964), 189-91
Marshall, Ann, 'Griffith Taylor, 1880-1963', *Geogr.
    Rev.*, vol 53, (1964), 427-9
Rose, John K., 'Griffith Taylor, 1880-1963', *Ann.
    Assoc. Am. Geogr.*, vol 54, (1964), 622-9
Tomkins, George S., *Griffith Taylor and Canadian
    geography*, Ph.D. thesis, University of Washington,
    (1966), 527p.
Andrews, John (ed), *Frontiers and men; a volume in
    memory of Griffith Taylor*, Melbourne, (1966), 186p.
Spate, Oskar H.K., 'Journeyman Taylor: some aspects
    of his work', *Austr. Geogr.*, vol 12, (1972), 115-22
Douglas, Ian, 'Frontiers of settlement in Australia--
    fifty years on', (6th. Griffith Taylor memorial
    lecture), *Austr. Geogr.*, vol 13, (1977), 297-305
Powell, Joseph M., 'The Bowman, Huntington and Taylor
    correspondence, 1928', *Austr. Geogr.*, vol 14, (1978),
    123-5
Powell, Joseph M., '1928: Griffith Taylor emigrates
    from Australia', *Geogr. Bull.*, vol 10, (1978), 5-13

*2. SELECTIVE AND THEMATIC BIBLIOGRAPHY OF WORKS BY
    T. GRIFFITH TAYLOR*

a. *Physical geography and geology*
1910 'The physiography of the proposed Federal Terri-
    tory at Canberra, Australia', *Bull. Bur. Meteorol.*,
    no 6, Melbourne, 18p.
1911 'The physiography of Eastern Australia', *Bull.
    Bur. Meteorol.*, no 8, Melbourne, 18p.
1913 *Climate and weather of Australia* (with H.A. Hunt
    and E.A. Quayle), Commonw. Bur. Meteorol., Mel-
    bourne, 93p.
1914 'Climate and weather', (with H.A. Hunt), in
    Oxford survey of the British Empire, edited by
    A.J. Herbertson and O.J.R. Howarth, Oxford, vol 5,
    91-139
--   'Mining and economic geology', in *Oxford survey of*

the British Empire, op. cit., vol 5, 216-67

1918 *Atlas of contour and rainfall maps of Australia,*
Advisory Council of Science and Industry, Melbourne

1920 *Australian Meteorology,* Oxford, 312p.

1927 'The topography of Australia', *Commonw. Yearb.* no 20,
Melbourne, 75-90

1928 'Glaciation in the South West Pacific', *Proc. Pan-
Pac. Sci. Congr.,* vol 2, Tokyo, 1819-25

1933 'The Australian environment', *Cambridge history of the
British Empire,* vol 7, 3-23

-- 'The soils of Australia in relation to topography and climate', *Geogr. Rev.,* vol 23, 108-13

1937 'Comparison of the American and Australian
deserts', *Econ. Geogr.,* vol 13, 260-8

1941 'The climates of Canada', *Can. Banker,* vol 49, 34-59

b. Antarctica

1913 'The Western journeys', in L. Huxley (ed), *Scott's
last expedition,* vol 2, London, 182-290

-- 'A résumé of the physiography and glacial geology
of Victoria Land, Antarctica', in L. Huxley (ed),
Scott's last expedition, vol 2, London, 416-29

1914 'Antarctica, the British sector', in *Oxford
survey of the British Empire,* edited by A.J.
Herbertson and O.J.R. Howarth, vol 5, 518-38

-- 'Physiography and glacial geology of East Antarctica', *Geogr. J.,* vol 44, 365-82, 452-67,
553-71

1916 *With Scott: the silver lining,* London, 464p.

1922 *British Antarctica (Terra Nova) Expedition
1910-13: the physiography of McMurdo Sound and
Granite Harbour Region,* London, 246p.

1928 'Climatic relations between Antarctica and Australia', in W.L.G. Joerg (ed), *Problems of
polar research,* New York, 285-99

1930 *Antarctic adventure and research,* New York, 244p.

c. Settlement controls and planning

1914 'Physical features and their effect on settlement', in *Oxford survey of the British Empire,*
vol 5, *Australia* ed. A.J. Herbertson and
O.J.R. Howarth, Oxford, 34-91

1915 'Climatic control of Australian production',
Bull. Bur. Meteorol., no 11, Melbourne, 32p.

1918 *The Australian Environment, Especially as Controlled by Rainfall - a regional study of the
topography, drainage, vegetation and settlement
and of the character and origin of the rains,*
Memoir No.1, Advisory Council of Science and
Industry, Melbourne, 188p.

-- 'Climatic factors influencing settlement in
Australia', *Commonw. Yearb.,* no 11, Melbourne,
84-101

1919 'The physiographic control of Australian exploration', *Geogr. J.,* vol 53, 172-92

-- 'The settlement of tropical Australia', *Geogr.
Rev.,* vol 8, 84-115

1920 'Agricultural climatology of Australia', *Q. J.
R. Meteorol. Soc.,* vol 46, 331-55

-- 'Nature versus the Australian', *Science and
Industry,* vol 2, 459-72

1922 'The distribution of future white settlement,
a world survey based on physiographic data',
Geogr. Rev., vol 12, 375-402

1924 'Geography and Australian national problems',
Australasian Assoc. Adv. Sci., Rep. of 16th
Meeting, Wellington, 433-87

1926 'The frontiers of settlement in Australia',
Geogr. Rev., vol 16, 1-25

1928 'The status of the Australian States, a study
of fundamental geographical controls', *Austr.
Geogr.,* vol 1, 7-28

1930 'Agricultural regions of Australia', *Econ. Geogr.,*
vol 6, 109-34, 213-42

-- 'The control of settlement in Australia by geo-
grahical factors', *Proc. 6th. Sess. Inst. Int.
Relations,* vol 6, Berkeley, 207-18

1932 'The pioneer belts of Australia', in *Pioneer Settlement, Am. Geogr. Soc. Spec. Publ.,* no. 14, New York,
360-91

-- 'The inner arid limits of economic settlement in
Australia', *Scott. Geogr. Mag.,* vol 48, 65-78

1937 'The possibilities of settlement in Australia',
in I. Bowman (ed), *Limits of land settlement,*
New York, 195-227

1940 *Australia: a study of warm environments and their
effect on British settlement,* London, 455p.

1942 'British Columbia, a study in topographic control',
Geogr. Rev., vol 32, 372-402

1946 'Future population in Canada - a study in technique', *Econ. Geogr.,* vol 22, 67-74

-- 'Parallels in Soviet and Canadian settlement',
Int. J., vol 1, 144-58

d. Cultural and political geography

1919 'Climatic cycles and evolution', *Geogr. Rev.,*
vol 8, 84-115

1921 'The evolution and distribution of race, culture
and language', *Geogr. Rev.,* vol 11, 54-119

1927 *Environment and race; a study of evolution,
migration, settlement and status of the races of
man,* London, 354p.

1929 'Racial migration zones', *Hum. Biol.,* vol 2, 34-62

1931 'The Nordic and Alpine races and their kin, a
study of ethnological trends', *Am. J. Sociol.,*
vol 37, 67-81

1933 *Atlas of environment and race,* Chicago, 110 maps
and diagrams

1936 *Environment and nation: geographical factors in
the cultural and political history of Europe,*
Toronto, 571 p.

1937 *Environment, race and migration,* Toronto, 483p.

1946 *Our evolving civilization: an introduction to
geopacifics,* Toronto, 370p.

e. Regional geography

1911 *New South Wales: historical, physiographical and
economic,* with A.W. Jose and W.G. Woolnough,
Melbourne, 372 p.

-- *Australia in its physiographic and economic
aspects* (ed A.J. Herbertson), Oxford, 256p.

1914 *A Geography of Australasia* (ed A.J. Herbertson),
Oxford, 176 p.

-- 'Evolution of a capital, a physiographic study
of the foundation of Canberra, Australia',
Geogr. J., vol 48, 378-95, 536-54

1916 *The world and Australasia*, with O.J.R. Howarth
     (ed A.J. Herbertson), Oxford, 423p.
1931 *Australia, a geography reader*, Chicago, 440p.
1946 *Newfoundland: a study of settlement with maps
     and illustrations*, Can. Instit. Int. Aff.,
     Toronto, 32p.
1947 *Canada. A study of cool continental environments
     and their effects on British and French
     settlement*, London, 524p.

f. *Urban geography*
1942 'Environment, village and city: a genetic
     approach to urban geography, with some
     reference to possibilism', *Ann. Assoc. Am.
     Geogr.*, vol 32, 1-67
1949 *Urban geography. A study of sites, evolution,
     pattern and classification in villages, towns
     and cities*, London, 439p.

g. *Education and philosophy*
1925 *The geographical laboratory*, with Dorothy Taylor,
     Sydney, 78p., also 1938, Toronto, 107p.
1935 'Geography – the correlative science', *Can. J.
     Econ. Polit. Sci.*, vol 1, 535-50
1938 'Correlations and culture – a study in technique',
     Brit. Assoc. Adv. Sci., Rep. Ann. Meet. Cambridge,
     103-38. Other versions published in *Nature*,
     vol 142, 737-41; *Scott. Geogr. Mag.*, vol 54,
     321-44; *Pan-Am. Geol.*, vol 70, 241-62, and vol 71,
     81-106
1942 'The role of geography', in F.L. Burdette (ed),
     Education for citizen responsibities, Princeton,
     44-61
1950 (ed) and 1957 *Geography in the twentieth century*,
     London, 674p.

3. *ARCHIVAL SOURCES*
The known papers of Griffith Taylor extend to approximately 40 linear feet.   The most extensive and representative collection is held by the National Library of
Australia in Canberra (26 linear feet), but the Department of Geography of the University of New England,
Armidale, New South Wales, retains much of his original
Antarctic material, including some diaries, letters and
illustrations.   In addition, Cambridge University holds
his lengthy unpublished manuscript dealing with his
experience there.   The author is indebted for unstinted
help to Mr. Alan Ives, librarian of the Australian
Archives, Canberra.

*Dr. J.M. Powell is Reader in Geography at Monash
University, Clayton, Victoria, Australia.*

DATES	LIFE AND CAREER	ACTIVITIES, TRAVEL, FIELDWORK	PUBLICATIONS	CONTEMPORARY EVENTS AND PUBLICATIONS
1880	Born in Walthamstow, England, 1 December			
1881	Lived in Serbia (to 1884) then in north of England			International Polar Year 1881-2. Death of Charles Darwin and George Perkins Marsh (both 1882)
1887				H.J. Mackinder appointed Reader in Geography, Oxford University
1888-91			Schoolboy pieces published in *Manchester Weekly Times*	
1893	Arrived in Sydney, Australia, educated at private schools			
1896-7	At King's School, Sydney where he met Frank Debenham			
1898	Entered Correspondence Board, Treasury, New South Wales			
1899	Student at Sydney University	Association with Edgeworth David began; geological fieldwork in New South Wales		
1901-03		Continued fieldwork, some in mining areas with Douglas Mawson	Geological papers from 1903	Federation of Australia, 1 January 1901. First Scott expedition to Antarctic
1904	B.Sc. with Honours in Physics and Geology			
1905-06	Bachelor of Engineering (1905); became Demonstrator in Geology, Sydney University	Fieldwork extended to Victoria, South Australia and Great Barrier Reef		A.J. Herbertson's paper, 'The major natural regions of the world', *Geogr. J.* vol 25 (1905) 300-10
1907	Gave the first formal lectures in geography at Sydney University; awarded 1851 Science Research Scholarship to Cambridge University	Began correspondence with W.M. Davis; travelled to England via New Zealand, Fiji, Hawaii, U.S.A. and Canada	Geological papers included two on coral reefs.	First American doctorate in geography awarded at Chicago University; publication of E. Huntington, *The Pulse of Asia*, London, Boston, New York
1907-08	At Emmanuel College, Cambridge University	Fieldwork with W.M. Davis in the Alps and Snowdonia; attended Dublin meeting, British Association for the Advancement of Science		Shackleton's successful Antarctic expedition; R.de Courcy Ward's *Climate, considered especially in relation to man* published
1909		Further fieldwork in Alps, Riviera and northern Europe		Peary reached North Pole; publications of *Geographical Essays* by W.M. Davis

DATES	LIFE AND CAREER	ACTIVITIES, TRAVEL, FIELDWORK	PUBLICATIONS	CONTEMPORARY EVENTS AND PUBLICATIONS
1910	B.A. (Cantab.); Fellow of the Geological Society; left Lyttelton, New Zealand, on Scott's *Terra Nova* as a representative of the Commonwealth Weather Service in the Antarctic	Survey of the site of Canberra with Edgeworth David and fieldwork in New Zealand with F. Debenham and Charles Wright	Paper on 'Physiography of Canberra', *Bull. Bur. Meteorol.* no. 6, Melbourne	
1910–12	On return from Antarctic resumed work as Physiographer with the Commonwealth Weather Service, Melbourne	Further fieldwork in Canberra area; on Scott's second expedition acted as geologist and leader of western parties	First general works on Australia published by Taylor and collaborators	Amundsen reached South Pole, 1911. E.C. Semple; *Influences o{f} Geographic Environment*, Chicago, 1911
1913	Engaged in collating and editing material on Antarctic Research under the auspices of the Royal Geographical Society and the British Museum	Travelled to England via Ceylon	Part author of *Climate and Weather of Australia*, Commonw. Bur. Meterol., Melbourne, and wrote on the Scott expedition	
1914	Married to Doris Priestley, 8 July in Melbourne, taught meteorology to air cadets	Lectured on Antarctic in South Africa; attended British Association meeting in Australia	*A geography of Australasia*, Oxford; four chapters in *Oxford Survey of the British Empire*. Contributions to *Handbook* of British Association meeting	Outbreak of 1914-18 war
1915			'Climatic control of Australian production', *Bull Bur. Meteorol.* no. 11, Melbourne	E. Huntington, *Civilizatio{n} and Climate*, New Haven
1916	D.Sc. Sydney University, elected adviser to Faculty of Science, Melbourne University		*With Scott: the silver lining*, London; *The World and Australasia*, Oxford	
1917	Attached to Geology Department, as external lecturer in geography (physiography); Thomson Gold Medal, Royal Geographical Society of Australasia (Queensland) for tropical research; his views challenged by J.W. Gregory	Correspondence with E. Huntington began	Further papers on Antarctic and on settlement of Australian tropics	
1918			*The Australian Environment*, Melbourne; Atlas of *Contour and Rainfall maps*, Melbourne	

DATES	LIFE AND CAREER	ACTIVITIES, TRAVEL, FIELDWORK	PUBLICATIONS	CONTEMPORARY EVENTS AND PUBLICATIONS
1919	Appointed to four Commissions, on agriculture and meteorology, solar radiation, aerial navigation and upper air research, of the International Meteorological Congress, Paris	Fieldwork in Tasmania and arid areas of South Australia	Papers including the first of several in the *Geographical Review*, New York	Isaiah Bowman, *The New World*, New York and Chicago
1920		Fieldwork in East Indies; correspondence with Isaiah Bowman began	*Australian Meteorology*, Oxford	H.G. Wells, *Outline of History*, London
1921	Became Associate Professor in charge of the newly founded department of geography at Sydney University; his textbook banned by the education authorities in Western Australia	Attended meeting of Australasian Association for the Advancement of Science, Melbourne and became founder member of the Australian National Research Council, representing meteorology	'The evolution and distribution of race, culture and language', *Geogr. Rev.* vol 11 (1921) 54-119	O.E. Baker's work on physical controls of agriculture in U.S.A. became well known
1923	The racial and settlement theories of Taylor became well known in Australia	Further fieldwork in tropical Australia	*Physiography of McMurdo Sound and Granite Harbour*, London	Death of Shackleton
1924		Fieldwork in arid Western Australia	Papers on Australian climates	E.Huntington, *Character of Races*
1925			Articles on arid Australia in *Sydney Morning Herald*	R.de Courcy Ward, *The Climates of the United States*
1926		Attended Pan-Pacific Science Conference, Tokyo; fieldwork in China, Japan, Korea, Philippines		
1927	First President, Geographical Society of New South Wales; joint editor of *Australian Geographer*	At Honolulu meeting of Institute of Pacific Affairs	*Environment and race*, London	
1928	Resigned his post at Sydney University	Fieldwork in Egypt on the way to England		First volume of the *Australian Geographer* appeared
1929	Became professor of geography at Chicago University	Lectured at Toronto and Clark Universities		
1930		Fieldwork in Panama and Colombia for American Geographical Society	*Antarctic Adventure and Research*, New York	
1931	Attended International Geographical Congress, Paris; Vice-President of Section E, British Association for the Advancement of Science, London meeting	Fieldwork in Scandinavia, western and central Europe	*Australia, a geography reader*, Chicago; papers on Colombia and on Nordic and Alpine races	I. Bowman, *The Pioneer Fringe*, New York

DATES	LIFE AND CAREER	ACTIVITIES, TRAVEL, FIELDWORK	PUBLICATIONS	CONTEMPORARY EVENTS AND PUBLICATIONS
1932		Fieldwork in the arid areas of southwest U.S.A.	More papers on arid Australia and one on geographers as national planners	Joerg, W.L. (ed), *Pioneer Settlement*, New York
1933			*Atlas of Environment and Race*, Chicago	
1934		Fieldwork in the northwest of U.S.A.		Death of W.M. Davis and also of Sir Edgeworth David. First volumes of A.J. Toynbee, *A study of history*, London
1935	Foundation professor of geography, Toronto University; adviser to Admiral Byrd's projected second expedition to Antarctica	Further work in the southeast of U.S.A. and in eastern Canada		
1936–38	Establishment and rapid growth of the department in Toronto with Donald Putman and George Tatham as main colleagues and Andrew Clark as first M.A. in 1938; President, Section E, British Association, Cambridge meeting, 1938	With Andrew Clark, fieldwork throughout Canada, in Western and Mediterranean Europe, also in northwest Africa	*Environment and Nation: geographical factors in the cultural and political history of Europe*, Toronto (1936); *Environment, race and migration*, Toronto (1937)	Bowman, I. (ed.), *Limits of Land Settlement*, New York
1939	Elected Vice-President, Association of American Geographers, Harvard meeting			Hartshorne, R, *The nature of geography*, Cambridge, Mass. Outbreak of war
1940	Introduction of Honours courses in geography, Toronto University; President, Association of American Geographers (first non-American holder of this office)		*Australia: a study of warm environments and their effect on British settlement*, London, 455p.; several articles on Europe, with work on Canada	
1941	Attended New York meeting of the Association of American Geographers	Fieldwork in Canada, especially in British Columbia		
1942	Fellow of Royal Society of Canada	Fieldwork in east Canada		
1944	Delivered Messenger Lectures on Civilization, Cornell University			
1945		Worked in Newfoundland		Huntington, E., *Mainsprings of Civilization*, New York. End of war

DATES	LIFE AND CAREER	ACTIVITIES, TRAVEL, FIELDWORK	PUBLICATIONS	CONTEMPORARY EVENTS AND PUBLICATIONS
1946			*Our evolving civilization: an introduction to geopacifics*, Toronto; *Newfoundland*, Toronto	
1947			*Canada : a study of cool continental environments and their effects on British and French settlements*, London	Deaths of H.J. Mackinder and of E. Huntington
1948	Adviser to the Interim Council of the Australian National University. Visited the campuses of Australian universities	On the way to Australia, fieldwork in Hawaii and Fiji		
1949			*Urban Geography*, London	Clark, A.H., *The invasion of New Zealand by people, plants and animals*
1950	Lecture tour of British universities arranged by British Council		*Geography in the twentieth century* (ed.), London	
1951	Retired and became Emeritus Professor of Toronto University. Honorary President of the new Canadian Association of Geographers			
1952	Returned to Sydney and became a council member of the Royal Society of New South Wales			
1954	Elected to Australian Academy of Science; President of Section P of the Australian and New Zealand Association for the Advancement of Science, Wellington meeting			James, P.E. and Jones, C.F., (eds.), *American geography: inventory and prospect*, Syracuse
1958			*Journeyman Taylor* (ed. A.A. MacGregor), London	
1959	First president, Institute of Australian Geographers			Hartshorne, R., *Perspective on the nature of geography*, Chicago
1961	Medal, Royal Society of New South Wales; honorary D.Litt., Sydney University			Griffith Taylor School of Geography opened at University of New England, Armidale, N.S.W.
1963	Died in Sydney, 5 November		Last papers on 'Probable disintegration of Antarctica', *Geogr. J.* 129 (1963) 190-1; 'Geographers and world peace: a plea for geopacifics', *Austr. Geogr. Stud.*, vol 1 (1963) 3-17	Publication of the first volume of *Australian Geographical Studies*

Alexey Andreyevich Tillo

1839–1900

I. A. FEDOSSEYEV

Alexey Andreyevich Tillo has gone down in the history of science as a Russian geographer, geophysicist, geodesist, and cartographer. He is known for his works on the hypsometry of Russia, and organized and conducted an important expedition to explore the sources of the major rivers in the European part of that country. He also researched the properties of magnetism.

1. EDUCATION, LIFE AND WORK

A. Tillo was born on 25 November 1839, into the family of a railway engineer. His father's ancestors were French Huguenots. When still a boy Alexey showed much zeal for learning and was fond of books, qualities which he inherited from his father, a well-educated man.

In 1849 the Corps of cadets, a special school for the children of noblemen and officers, was opened in Kiev and Tillo became one of its first students. In 1856 he was transferred to the St Petersburg Corps of cadets and graduated from it in 1859. In 1862 Tillo graduated from the St Petersburg Artillery Academy and then studied geodesy for two years at the Academy of General Headquarters. After this he spent two years improving his geodetic and astronomical knowledge under O.W. Struve at the Observatory of Pulkovo. From 1867–71 he headed the Military Topographical Department in the Orenburg Military District and from 1871–74 served with the troops before working at the General Headquarters of Russia. From 1879–83,

while tutor to a member of the Tzar's family who travelled abroad to study, Tillo took the opportunity to attend the lectures delivered by prominent scholars in Germany, and the degree of Doctor of Philosophy was conferred on him at Leipzig University. He was Chief of Corps Staff and Commander of a Division from 1883–99, and in 1894 received the rank of Lieutenant General.

An important stage in Tillo's life was his active participation in the work done by the Russian Geographical Society. In 1884 he was elected a member of the society council; in 1889 he became Chairman of the Mathematical Geography department of the society and in 1898 its Assistant Chairman. He was also a corresponding member of the Academy of Sciences in Paris (from 1892), and the Academy of Sciences in St Petersburg (from 1894).

Sent as a delegate to a number of international geographical congresses, Tillo was elected Honorary Vice-President at the 7th Congress held in Berlin in 1899. Here he moved a proposal to establish the International Cartographic Association, which was approved by the rest of the participants in the Congress. Besides Russian, which was his native tongue, he knew French, German, English, and Italian. Tillo expected a great deal of himself and his colleagues and yet was a kindly and considerate man.

2. SCIENTIFIC AND GEOGRAPHICAL WORK

Although Tillo's career is associated with military service he will be remembered for his great contribution to science. His scientific career dates from 1866, when he translated into Russian the geodetic works of the German scholars K.F. Gauss, O.F. Bessel and P.A. Hansen. When translating one of the papers written by K.F. Gauss, A. Tillo supplemented it with a table made up by himself.

From 1867-71, Tillo headed a team which determined astronomically the coordinates for a number of base stations in the Orenburg district. He introduced a new triangulation for this district and supervised a topographical survey covering an area of about 36,000 sq km. Later the Geographical Society charged him with the task of levelling the land stretching from the Aral Sea to the Caspian Sea, and while fulfilling this assignment he found that in the summer of 1874 the level of the Aral Sea was 74m higher than that of the Caspian Sea. Tillo was also one of the sponsors and organizers of the large-scale levelling of the Siberian territory from the Urals to Lake Baikal, which was carried out by the Geographical Society in 1875-7.

In 1885 he completed the measuring of the absolute heights for Lake Ladoga, Lake Onega, and Lake Ilmen, arriving at results which differed considerably from those to be found in other published sources. For example, according to E. Reclus the absolute height value for Lake Ladoga was 59 ft, Lake Onega 237 ft, and Lake Ilmen 157 ft, whereas Tillo's values were 16.5 ft, 114.9 ft and 58 ft respectively. Tillo drew up the levelling description for the rivers and a catalogue listing the absolute levels of the water basins in European Russia. He did, moreover, verify on the available maps the length of 155 major rivers in European Russia totalling 77,000 km and made corrections to data previously obtained.

His main work in the field of hypsometry was an orographic map for European Russia for which he used more than 51,000 spot elevations. This work, completed in 1889, was the first correct representation of the relief of European Russia. Before Tillo made his map it was believed that the territory of European Russia was crossed by two ridges stretching its width, those of the Urals-Baltic and Urals-Carpathians. In fact, it turned out that the territory has two meridian highlands which Tillo named the Middle-Russian and Near-Volga (Privolzhsk) Highlands.

Also in 1889 Tillo published his paper on the mean height of the continents and the mean height of the oceans. In particular, he showed that the height of continents and of oceans grows as one moves away from the equator, reaching the largest values within the latitudes of 30-40°, and dropping again towards the poles.

In addition to his hypsometric works, Tillo made a considerable contribution to science by studying the elements of terrestrial magnetism over the area of European Russia, which he began in 1869. He wrote more than thirty works on the problems of terrestrial magnetism. In 1882 he published what were the earliest maps to describe the distribution of terrestrial magnetism as well as magnetic anomalies throughout European Russia.

Tillo's contribution to the exploration of the drainage system in Russia, apart from the above-mentioned works on rivers and lakes, included a map, issued in 1897, which showed basins of inland water routes with data on the basins of 217 of Russia's rivers. Another major contribution came

from the special and very fruitful expedition in 1894 led by him to explore the sources of the major rivers in European Russia. Tillo's *Atlas of rainfall distribution over the river basins in European Russia* (1897) was based on data obtained on the expedition and he later represented Russia on the International Commission making the World Geographical Map to a scale of 1:1,000,000.

Tillo, a very talented and extremely hard-working man, did so much for science that his name can justly be listed among the most prominent Russian geographers in the second half of the nineteenth century.

Bibliography and Sources

1. REFERENCES ON A.A. TILLO
Grigoryev, A.V., 'A.A. Tillo', *Rep. Russ. Geogr. Soc.*, St Petersburg, (1900)
Berg, L.S., 'Alexey Andreyevich Tillo', *Proc. All-Union Geogr. Soc.*, vol 82, no 3, (1950), 113-23
Burmistrov, G.A., 'A.A. Tillo', *Trans. Moscow Inst. Geod. Aerial Photogr. Cartogr.*, no 15, (1953), 49-60
Novokshankova, Z.K., *Alexey Andreyevich Tillo*, Moscow, (1961), 120 p. (with the most complete bibliography available)

2. PRINCIPAL WORKS OF A.A. TILLO
1870 'Report on the activities of the Military Topographical Department of the Orenburg military district for 1868', *Trans. Orenburg Dept. Russ. Geogr. Soc.*, no 1, Kazan, 259-70
1873 'Astronomical fixation of geographical locations in the Orenburg district carried out in 1867-71. Part I. The results obtained by the expeditions working between the towns of Orsk and Kazalinsk (with plans of the surroundings in the vicinity of the astrofixes and Fort I)', *Trans. Military Topogr. Dept. GHQ*, part 33, 137-91
1875 'Astronomical fixation of geographical localities in the Orenburg district carried out in 1867-1871. Part II. The results obtained by the Expedition of 1870 in the Ural and Turgay regions. Part III. The Expedition of 1871 at the fortress of Akatube', *Trans. Military Topogr. Dept. GHQ*, part 34, 105-78
1877 *Description of the Aral-Caspian levellings carried out in 1874 according to the order of the Imperial Russian Geographical Society and its Orenburg department*, St Petersburg, 42 p.
1879 'On terrestrial magnetism in European Russia', *Proc. Russ. Geogr. Soc.*, vol 15, no 4, 279-88
—— 'On the works carried out for grade measurements in Europe by the end of 1877 and on the latest conclusions arrived at by British geodesists on the form of the Earth after their work in India', *Proc. Russ. Geogr. Soc.*, vol 15, no 2, 57-65

1881 *An attempt to compile a description of levellings in the Russian Empire. Materials on the hypsometry of Russia, with the Atlas of longitudinal profiles.* Scales: for horizontal distances – 1:420,000, ten Russian versts to one English inch; for heights 1:1,680, 20 sazhens to one inch. Section I.

–– *The longitudinal profiles for the railways to be constructed by the Ministry of Communications,* St Petersburg, 1881, 48 p. with drawings. Section II. *The longitudinal profiles of the railways constructed,* St Petersburg, 43 p. with drawings. Section III. *The longitudinal profiles of the paved high roads,* St Petersburg, 1882, 16 p. with drawings. Section IV.

–– *The longitudinal profiles of rivers and canals,* St Petersburg, 1882, 24 p. with drawings.

1882 'The collected description of levellings for European Russia', *Proc. Russ. Geogr. Soc.,* vol 18, no 5, 281-95

–– 'On geographical science in Germany. Considerations on the German Geographical Society', *Collected Works of the Navy,* vol 190, no 6,, pp. 53-73.

1883 'Investigations into the geographical distribution of magnetism and secular magnetic variations and dips over the territory of European Russia (with 4 maps)', *Meteorol. Collected Works, Acad. Sci.,* vol 8, no 2, 82 p.

–– 'On the length of the rivers of European Russia', *Proc. Russ. Geogr. Soc.,* vol 19, no 3, 133-65

1884 *Map of the heights in European Russia. Scale: 1:2,520,000,* St Petersburg.

1888 *Map of the length and downfall of the rivers in European Russia,* St Petersburg

1889 *Hypsometric map of European Russia. An attempt to present the structure of the surface in European Russia,* St Petersburg

1890 'Distribution of the atmospheric pressure over the territory of the Russian Empire and the Asian continent, proceeding from the observations of 1836-85 (with the atlas of 69 maps)', *Proc. Russ. Geogr. Soc.,* vol 21, 308 p.

–– 'The orography of European Russia based upon the hypsometric map (with 2 maps and 1 table)', *Trans. 8th Congr. Russ. Nat. Physicians,* St Petersburg, 28 December 1899 – 7 January 1890, 85-96

1891 'On terrestrial magnetism. The lectures delivered by Major General A. Tillo at the Navy Assembly in Kronstadt on 4 March 1891', *Collected works of the Navy,* vol 219, 1-22

1892 'The collected description of the river levellings, their downfall, and the catalogue of absolute altitudes of water levels in European Russia. Documents on hypsometry in the Russian Empire', *Railway Communications Journal,* supplement for April-May 1892, 131 p.

1897 *The map of basins of inland water routes in European Russia with notes on the meteorological and water-measuring stations. Scale: 1:2,520,000.* Department for Highways and Waterways attached to the Ministry of Communications. St Petersburg, 6 sheets

–– *Explanatory notes to the Map of inland water routes in European Russia, with tables showing the areas of river basins and the lists of meteorological and water-measuring stations,* St Petersburg, 26 p.

–– *Atlas of the Distribution of monthly and yearly rainfalls over the river basins in European Russia based upon twenty years' observations between 1871-1890,* Transactions of the Expedition which explored the sources of European Russia's major rivers, no 16, St Petersburg (6 pages of Explanatory notes, 14 sheets of maps).

Professor I.A. Fedosseyev is on the staff of the Institute of the History of the Natural Sciences, Moscow.

DATES	LIFE AND CAREER	ACTIVITIES, TRAVEL, FIELDWORK	PUBLICATIONS	CONTEMPORARY EVENTS AND PUBLICATIONS
1839	Born 25 November in Kiev			
1845				Russian Geographical Society organized
1849	Studied at Kiev Corps of cadets and of Petersburg Corps of cadets (to 1859)			
1856				P.P. Semenov's exploration in Asia (to 1858)
1860	Studied at the Artillery Academy in St Petersburg (to 1862)			
1862	Studied at Geodetic Dept. of the General Headquarters Academy in St Petersburg (to 1864)			
1867	Supervision of astronomical and geodetic works in the Orenburg military district (to 1871)	Carried out astronomical observations on The northeastern coast of the Aral Sea and in the lower part of the Sydarya river (to 1868)		
1871	Battalion commander later commander of a regiment in St Petersburg and Kronstadt (to 1874)			
1874	General Headquarters officer in Petersburg (to 1883)	Works on the levelling between the Aral Sea and the Caspian Sea	Description of the Aral-Caspian levelling done according to the instructions of the Imperial Russian Geographical Society in 1874 (and of its Orenburg dept. in 1877)	
1875				Establishment of Navigation Inventory Commission under the Ministry of Communications which explored all Russia's major rivers (to 1884)
1878		Treatment of the data for all the levellings carried out by the Ministry of Communications		

DATES	LIFE AND CAREER	ACTIVITIES, TRAVEL, FIELDWORK	PUBLICATIONS	CONTEMPORARY EVENTS AND PUBLICATIONS
1883	Held the commanding ranks in the Army (Chief of Corps Staff Commander of Division) (to 1899)			
1884			*Map of heights in European Russia*	A.I. Voyeikov, *Climates of the globe, with particular reference to Russia*
1888			*Map of the length and downfall of the rivers in European Russia*	
1889		Compiled the orographic map of European Russia. Supervised expedition to explore the sources of European Russia's major rivers	*The mean height of land and the mean height of sea*	
1890		Treatment of the data on the atmospheric pressure in the river basins of European Russia	'Distribution of atmospheric pressure over the territory of the Russian Empire and the Asian continent'; 'Orography of European Russia based upon the hypsometric map'	
1891			'On terrestrial magnetism'	
1897			*Atlas of rainfall distribution over the river basins in European Russia*	
1900	Died 11 January in St Petersburg (Leningrad)			

Zachris Topelius
1818–1898

W. R. MEAD

1. EDUCATION, LIFE AND WORK

Zachris Topelius, the first Finn to identify geography as an independent discipline, was born on
14 January 1818 in the small Ostrobothnian sea port
of Uusi Kaarlepyy (in Swedish, Gamla Karleby). He
was educated at the Latin school at Oulu (Uleåborg).
On 8 September 1828 Topelius noted in his diary that
he 'began to read geography' and on 14 December he
first 'drew a map'. The geographical background of
Finland permeates his writing. He was a student at
the University of Helsinki (then the Imperial Alexander University of the Grand Duchy of Finland)
between 1833 and 1847, and held an editorial appointment with the newspaper *Helsingfors Tidningar*, to
which he contributed regularly between 1841 and 1860.

In 1854 he gave his inaugural lecture, 'On
Finland's Geographical Position', on his election to
a personal chair in history. Between 1854 and 1858
he delivered the first extended course of lectures in
geography to be given at Finland's university. He
was appointed to the chair in Finnish, Russian and
Scandinavian history in 1863, and continued to
lecture with a strong geographical component in his
courses until 1873. Topelius's travels in England
and mainland Europe were recorded in perceptive
diaries. In 1875 he became rector of the University
of Helsinki (Helsingfors). As an educationist, he
held strong religious beliefs and his last publication, *Leaves from my book of thoughts*, was a sketch
of his faith. He died at Sipoo (Sibbo) on 12 March
1898.

2. SCIENTIFIC IDEAS AND GEOGRAPHICAL THOUGHT

Topelius's geographical contribution, unprinted and
largely unknown, dominates the 3,500 pages of his
manuscript lectures. The landscape of his home
country, for which he was already using the Finnish
name Suomi, was intimately known as a result of his
regular journeys by horse and cart through it. The
shape and nature of Finland were constantly in his
mind's eye. He taught 'with map in hand' and wall
map behind him, using C.W. Gyldén's orographic map of
1848 or M.K. Broström's relief map of 1858-9. He
began with the local, and proceeded to the universal.
For him, his home river valley was the apotheosis of
all river valleys, and he evolved a system of interpreting his home and other countries on a river
valley basis. He extended the 'expressions of their
owner's personality' as displayed in the features of
'a dwelling house, a farm, a ploughed field or an
orchard' to the appreciation of a nation. Topelius
was fundamentally an evironmentalist (often a romantic
one), conscious of the influence that his early geographical environment had had upon him, and attribu-
ting the behaviour of nations and states to their
geographical location and natural conditions. He
believed that 'physical circumstance and natural
process stand in the closest community'. For him,
the keys to human history were geographical setting,
national circumstances and the technical ability of
people to make use of them. Following in the foot-
steps of the Swedish historian E.C. Geijer (1783-
1847), he preached that 'geography lies behind
history' and that the two are inseparable, 'the one
illumines and gives authority to the other'.

For Topelius, the purpose of geography was to

gather together and present as a unity the varied
information about the natural world and man that the
sciences had discovered (Nyberg, p. 335).   As
a unifying subject, he presented geography as 'the
link between natural history and human history,
(holding) the key to both of them'.   No one was more
explicit about man-land relationships: 'the land is
a natural influence working on the people while the
people are an intelligent force working on the land
and striving to release it from its restraints.'
The discipline had other qualities.   It was dynamic,
for 'no subject can stand still while the materials
of its study are continually developing.'   It was
explanatory, for 'we should not entertain ourselves
with facts alone, without seeking their causes.'
And, being so central to explanation, geography was
more than 'a tributary science' (his own phrase)
dependent upon related diciplines.   It was 'an index
to the great book of nature and life.'   It had both
local and universal contexts.

Topelius used his geographical ideas for national
purposes and they were among the influences that made
him one of the founding fathers of independent Finland.   He sought to educate the Finns about their
land, to explain how they came to be what they were
and how they might improve their lot.   He made them
look at their country with new eyes, from new angles
and always in the dimension of historical time.   He
sought a reassessment of the inhibiting winter, as in
his *Twilight tales*, where he underlined the positive
side of the account.   For artist and naturalist
alike, he taught a new appreciation of the quality
both of light and of darkness.

Drawn to the tables of Finland's first official
statistician, Gabriel Rein (1800-1867), Topelius
aimed to present the economic geography of Finland in
the light of them.   They enabled him to stress both
the shortcomings of his country, especially in silvi-
culture, and its potentialities, not least in agriculture.   Given regular statistical returns of the
changing extent of cropland and meadowland, and of
the changing numbers of population and livestock, he
pointed out that it should be possible to avoid many
of the catastrophic shortages that Finland had suffered in the past.   Topelius was also able to use
statistics to impress the need for improved forest
management, suggesting that 'a quarter of Finland has
been rendered an unproductive wilderness' by the
wasteful use of woodland for tar burning, charcoal
burning and swidden (burn-beating).   The consequences
of its destruction were thought to be all the more
serious because of the presumed significance of
timber cover for tempering climate in high latitudes.

He was also one of the first Finns to talk quan-
titatively about the extent of bogland in Finland,
and to act as an active propagandist for land drainage.   In the light of his concern for figures, he
estimated that, if Finland's resources were properly
developed, it could support a population of 8 millions. (It was 1.5 millions in his day and is 4.7
millions today.)   Furthermore, as a proto-social
scientist conscious of the underdevelopment and
underemployment about him, he asserted that 'if there
is an unemployed proletariat it is not nature's
fault, but out own.'

Geography had placed Finland between two great
powers, Russia and Sweden, so that it had always been
their buffer and battleground.   Topelius was a
geopolitician before his Swedish *confrère* Rudolf
Kjellén (1864-1920) had invented the word, and he
regarded states and nations as organisms.   For the
Finnish nation to reach organic maturity, he believed
that 'it must be neutral, independent and free to
look after its own interests.'   It would then be
able to be a bridge between Russia and Scandinavia --
a thesis maintained by Finland's two post-war presi-
dents.   Looking at the map, Topelius also identified
two geographical concepts: Suomensaari, a greater
Finnic area, and Suomenniemi.   Suomensaari had the
Scandinavian pensinsula to its west and was separated
from Russia in the south-east by the line of the
Finnish Gulf, Lake Ladoga, Lake Onega and the White
Sea.   He saw it as a distinct isthmian area which
also had a distinctive ethnography because it was
occupied essentially by Finnic peoples.   (That
Finland might also embrace this area politically, was
an idea which moved Finnish activists during the
Second World War.)   Suomenniemi, the south-western
peninsula of Finland, at the meeting ground of the
Gulfs of Bothnia and of Finland and the Baltic Sea,
was a natural economic and commercial focus.   His
judgement was confirmed by the increasing concentration of population and economic activity there in the
mid twentieth century.

It was in Suomenniemi that the 'centuries' long
process of assimilation' of ideas and institutions
derived from western Europe took place.   Topelius
saw the course of all European history as cosmopolitan.
It was the story of 'how from being Asians /its
peoples/ have become Europeans'.   The differences
between the peoples and constituent states of Europe
lay 'only in the time and circumstances of their
conversion'.   Because of their location, the Finns
suffered a technical and cultural retardation.   This
was both good and bad: good, in that the country
absorbed outside ideas slowly and therefore retained
many elements of its unique culture; bad, in that
Finland, 'dragged into history by the scruff of its
neck', was slow to learn the need for, and meaning
of, history.

3. INFLUENCE AND SPREAD OF IDEAS

While Fridtjof Nansen (1861-1930) in Norway and Sven
Hedin (1865-1952) in Sweden caught the imagination of
prospective geographers, Zachris Topelius engaged the
attention of their Finnish counterparts by the
original way in which he presented his country.
In addition, he had a literary style which enabled
him to arouse and sustain interest among readers of
all ages.   The wide-ranging subject matter that he
employed always took off from or returned to the
land, its people and the relations between them.
Among his many contributions to the Finnish newspaper
Helsingfors Tidningar, those on the Saimaa canal (for
which the working papers exist among his manuscripts)
and on poverty in Helsinki (based on personal house-
to-house visits in its slum neighbourhoods) illustrate
the way in which he sought to inform public opinion in

the fields of what today would be called economic geography and social geography respectively. His serious poetry (such as *The ice break*) was shot through with a keen sense of physical geography, while he could also write schoolroom verse (such as *The map of Europe* and *How William learnt his geography*) to teach young readers elementary facts about the continent in which they lived.

As an academic, Topelius was the first Finn to lecture on the importance of documentary sources for building up an accurate picture of his country's social and economic past. He directed attention to the wealth of large-scale maps that several generations of surveyors had made for the purposes of land reorganization and taxation, to the population census (founded in 1749), and to the agricultural statistics prepared by C.C. Böcker (1786-1841), secretary of the Finnish Economic Society. While phenology had been stimulated in the eighteenth century by C. Linnaeus (1707-1778), Topelius was in a position to direct attention to the accumulated records. He turned to them for his interpretation of the causes and consequences of climatic change. As one who cultivated and fostered the art of observation, he was able to draw on a variety of personal experience to stimulate his students. No one could fail to respond to his appraisal of the isostatic behaviour of the Ostrobothnian coastlands. Through his reading, he strove to pass on new ideas and his lectures trace his conversion from a diluvial explanation of the glaciological features of Finland to an interpretation based upon the work of Agassiz. The Gulf Stream was a firmly established feature in his climatological explanations, and he introduced the concept of the isotherm to Finns. By the 1870s he was casting an eye over the distribution of wealth as well as population and resources. He believed that social problems had become of greater importance than international politics: 'The most important is the question of labour's relation to capital -- everything combines to concentrate capital in a few hands. ... [the most urgent question is] to emancipate labour from capital.'

As a teacher, Topelius aimed to inform the widest possible circle about the geographical circumstances of Finland. This is illustrated by three successive publications: *Finland framestalld i teckningar (Finland presented in drawings)*, a descriptive study of the provinces of Finland with 120 lithographs; *Boken om vårt land (The book about our country)*, a geography and history of Finland for the schoolroom and probably the single most influential book written in Finland; *En resa i Finland (A journey in Finland)*, an appreciation of the Finnish landscape illustrated with 36 steel engravings. As a pedagogue, he prepared the ground for the establishment of geography as an independent discipline in the University of Helsinki in 1890, and his former students were counted among the founder members of the Finnish Geographical Society in 1888. He lived to see the establishment of its research series, *Fennia*, in 1889, and the preparation of the *Atlas of Finland* (1899), the world's first national atlas. Among his last contributions were those to *Finland in the nineteenth century* (1894), a handsome folio volume,

published in English, French and German as well as the languages of Finland. It was a book in keeping with the objective of Topelius, to present an image of the Grand Duchy as a distinctive entity to the rest of the world.

Bibliography and Sources

REFERENCES ON ZACHRIS TOPELIUS
There is a printed bibliography of the principal published works of Zachris Topelius by B. Luneland-Grönroos, *Zachris Topelius tryckta skriver*, Helsingfors, 1954. There are two standard biographies, both in Swedish - V. Vasenius, *Zachris Topelius*, Helsingfors, 1912-33, 6 volumes and P. Nyberg, *Zachris Topelius*, Helsingfors, 1949. The substantial manuscript collection, of which there is a typescript catalogue (*Topeliuksen Kokoelma*), is in the University Library, Helsinki. This paper is based principally on the manuscripts of the lectures given between 1854 and 1873. There are two short contributions in Finnish on Topelius's geographical lectures 1854-8 by Sigrid Stenij, *Terra*, Helsingfors, 1937, 49, 1 and 2, cf. also W.R. Mead, 'Zachris Topelius', *Norsk Geografisk Tidsskrift*, 1968, 22, 89-100, and *The geographical tradition in Finland*, 1963, 20p. *Finland, A daughter of the sea* by Michael Jones (London, 1976) provides an illuminating account of the coastland on which Topelius was brought up - and employs a Topelian epithet in its title.

1848-52 *Finland framstalld i teckningar (Finland presented in drawings)*, Helsingfors, 260p.
1872-4 *En resa i Finland (A journey in Finland)*, Hesingfors, 180p.
1875 *Broken om vårt land (The book about our country)*, Helsingfors, 471p.
1894 *Finland in the nineteenth century*, Helsingfors, Introductory chapters, 5-82
1922 *Sjalfbiografiska anteckningar (Autobiographical notes)*, Helsingfors, 364p.

W.R. Mead is Professor of Geography at University College, London.

Index

The index is divided into four parts;

1. *PERSONAL NAMES* as far as possible are given in full for those mentioned in the text with the year of birth and of death.
2. *ORGANIZATIONS AND RELATED REFERENCES* is subdivided as in volume 2, that is under *Colleges*, *Institutes*, *Institutions*, *Museums*, *Official and Research Organizations*. *Scientific Congresses and Commissions*. *Societies*. and *Universities*.
3. *SUBJECTS* covers concepts, geographical theories and specific research.
4. *CUMULATIVE LIST OF BIOBIBLIO-GRAPHIES* indexes all the geographers studied in volumes 1, 2 and 3.

Page numbers in italic refer to the Bibliography and Sources sections of the biobibliographies and underlined numbers refer to the chronological tables.

1. *PERSONAL NAMES*

2. ORGANIZATIONS AND RELATED REFERENCES